W9-ASF-073

On the Side
of
the Apes

BOOKS BY EMILY HAHN

Hongkong Holiday
Raffles of Singapore
Miss Jill
England to Me
Purple Passage
Love Conquers Nothing
Chiang Kai-shek
Diamond
The Tiger House Party
China Only Yesterday
China to Me
Africa to Me: Person to Person
With Naked Foot
Romantic Rebels
Animal Gardens
Times and Places
On the Side of the Apes

On the Side of the Apes

by Emily Hahn

Thomas Y. Crowell Company

ESTABLISHED 1834 · NEW YORK

Jacket drawings by Joel H. Ito, *Primate News,* vol. 6, December 1968.

A substantial portion of this book originally appeared in *The New Yorker*.

 Published simultaneously in Canada by Fitzhenry & Whiteside Limited, Toronto.

Designed by Barbara Kohn Isaac

Manufactured in the United States of America

L.C. Card 78-146282
ISBN 0-690-59992-7

1 2 3 4 5 6 7 8 9 10

The author wishes to give special thanks to Roberta Yerkes Blanshard and James Augustine for their assistance in obtaining the illustrations.

Picture Credits

Yerkes Papers, Yale Medical Library: pp. 83–85; p. 86, *top;* pp. 87–89

Roberta Yerkes Blanshard: p. 86, *bottom*

Yerkes Regional Primate Research Center: pp. 90–91

Wisconsin Regional Primate Research Center: p. 92, *top* (Robert O. Sponholz); p. 92, *bottom* (Robert O. Dodsworth); p. 93, *top* (Jerry Hecht)

Washington Regional Primate Research Center: p. 93, *bottom left* (Betty L. Abel); p. 94, *left*

Oregon Regional Primate Research Center: p. 93, *bottom right,* p. 97 (Robert Reynolds); p. 95, *top right* (Barton L. Attebery); p. 95, *bottom right* (Herb Alden); p. 96

National Center for Primate Biology: p. 98

Delta Regional Primate Research Center: pp. 99–101 (E. W. Menzel, Jr.); p. 102 (Arthur J. Riopelle)

For Carola and Amanda

Contents

Aug. 24th, 1661: At the office all morning and did business; by and by we are called to Sir W. Batten's to see the strange creature that Captain Holmes hath brought home with him from Guiny; it is a great baboon, but so much like a man in most things, that though they say there is a species of them, yet I cannot believe but that it is a monster got of a man and a she-baboon. I do believe that it already understands much English, and I am of the mind it might be taught to speak or make signs.

—*The Diary of Samuel Pepys*

The question is this: Is man an ape or an angel? I, my lord, I am on the side of the angels.

—Benjamin Disraeli

I

The Dream

The director of the Primate Research Center spoke with unusual emphasis, I thought. "Some of us go in for virology; others are studying the cardiovascular system. We have enough behaviorists to stock all the universities in the country. Physiology, embryology—you name it; we study it. But all any member of the public asks me about is the possibility of breeding apes with people. I mean to say, nine out of ten have never even heard of the primate centers, but that one man, the tenth one, he's got that idea about primatology, and nothing seems to shake it. It's an obsession."

He shoved some articles around on his desk half-angrily. "I remember when I was first in charge of the old Yerkes Station at Orange Park near Jacksonville. The directors before me had ignored the public-relations angle because they thought it would interfere with their work, but I've always been a great believer in letting people know what goes on in institutions like ours. People resent mystery, especially when they suspect it's costing them in taxes, and I see their point. So I waded right in and asked the mayor of Jacksonville, the first time I met him, if he'd like to encourage people to come and see our place—tours for high school classes, that kind of thing. Well, he didn't seem to take to the idea, and finally he asked me plainly if we weren't crossbreeding apes with humans."

The director gave me a disgusted glance and sat back in his chair. "Of course I denied it, until after a while he seemed about half convinced. Anyway he invited me to give a talk to one of those Rotary lunches, and I did. I described our work and what we hope to get out of it for humanity, and so forth, and then I asked if they had any questions. They did, too. Every single question had something to do with crossbreeding apes and people.

"Well, I talked to that club until I was blue in the face, telling them we weren't attempting to do this and didn't intend to attempt it. It wouldn't work anyway, I said. I guess some of them finally believed me —anyway they stopped showing any interest whatever in Yerkes—but that mayor still wasn't quite sure, until I looked him in the eye and said, 'It can't be done. For one thing, no American woman would consent to take part in an experiment like that, would she?'

"You know, that really shut him up. He'd always thought in terms of man with female chimp, not male chimp and woman." The director paused for a second, in memory savoring his triumph; then his face fell. "I still get questions about it, though," he said.

In justice to the citizens of Jacksonville and all who share their fears of primatology, it should be stated that they didn't make up this delicious horror of an idea out of their own dark subconscious. The idea is a recurring one with many people outside the United States as well as inside. For example, Russian scientists, at a time when all of us knew somewhat less than we do today about chromosomes and species differences, gave it serious thought. I discovered this fact (and others as well) in a little book, *Monkeys for Science,* by Professor Boris Lapin, M.D., director of the Institute of Experimental Pathology and Therapy, USSR Academy of Medical Sciences, and Eman Fridman, chief of the Primate Information Center, of the same Institute of Experimental Pathology and Therapy. Their main theme in the book is the Russian primate center, or as they often call it, the monkey-breeding center, at Sukhumi on the Black Sea.

"Man's attention has long been attracted to the monkey—that grotesque copy of himself," the preface begins. "In his ignorance he endowed it with the most unbelievable characteristics. Scientific acquaintance with this animal was protracted over centuries. There was much that hampered recognition of the truth, but there is no force which can eternally or even for very long stand in the way of reason." After this cheerful expression of faith, the writers continue:

> There came a time when man, plucking up courage, declared loudly and with adequate grounds: this is our cousin; we are blood relations—we have similar forbears, our bodily structure is very similar and apparently we suffer from the same diseases.

> The monkey, our ancient kinsman, opened such opportunities before the researcher that today it is hard to conceive of the science of human disease without man's laboratory twin. Measles, poliomyelitis, typhoid fever, dysentery—models of these and many other diseases became available to the experimenter through the use of monkeys.

The book itself opens with a brief account of apes and monkeys in history, mentioning among other species the sacred hamadryad baboon of Egypt.

> Monkeys were deified and are still worshipped as divinities by Tibetans, Malayans and Hindus. In contrast, some other peoples (Arabs, Mexicans, etc.) contended monkeys were "evil spirits," "the devil's offspring," sinners whom God had condemned by turning into an ugly likeness of man. The Russian word *obezyana* for "monkey" is also an echo of such a belief, being derived from the Arabic words meaning "father of sin.". . . And still, even in the XIX century one could come across writings about the "mandrill devil" which "comes to town twice a week to collect tribute" and other fantastic details associated with monkeys.

Before the middle of the nineteenth century, the writers say, "the prominent Russian naturalist and natural historian K. Ber, whom F. Engels regarded as Darwin's predecessor and to whom Darwin himself referred in his works," had counted only 150 species of monkey in the whole world; another Russian, "the leading Soviet primatologist M. Nesturich, dates the shaping of primatology as a science only from the middle 50's of our century." There are various reasons for this delay: it was hard to catch monkeys and harder still to keep them alive after capture, but the largest impediment was public prejudice.

> The monkey had for many centuries been a "cursed" animal. Its anatomical similarity to man . . . seemed a reproach to the authors and advocates of the biblical version of the divine creation of man. . . . It can be well supposed that had it not been for the zeal of the Holy Inquisition which destroyed essays on man's kinship with the monkey, and oftentimes the authors along with the essays, much more information on this question would have come down to us from the Middle Ages. Getting a scientific knowledge of monkeys had a direct bearing on the fundamental problem of man's origin, and was thus in the focus of the bitter fight between idealism and materialism. The Inquisition sent Giordano Bruno and Lucilio Vanini to the stake; even in the XVIII century, ecclesiastics exerted pressure on the French naturalist Georges Buffon who classified monkeys and human beings as "quadrumanes" and "bimanes."

Here Lapin and Fridman lost me. Giordano Bruno—What had he to do with monkeys? He was a cosmologist. Then, reading the passage over, I realized that nobody had actually *said* these victims had gone to the stake for believing in man's relationship with monkeys; it was merely said that they had gone to the stake. Nor did the authors state outright that Buffon had suffered for lumping monkeys and us under one classification by calling monkeys four-handed and us two-handed. They merely hinted it.

Resuming my reading, I learned of Darwin and Thomas Huxley, and of what happened to their works in Tsarist Russia:

> And here is an example of how reactionaries fought materialism in science, in particular in primatology and anthropology. Huxley's book, published in St. Petersburg in 1864, was a tremendous success among the Russian public. A second edition had to be published in the same year. But the translation, which came out in Moscow this time, was stripped of all the author's conclusions which were revolutionary for the time. . . . In fear of the materialistic advance in science, Russian reactionaries committed outright falsification, deleting all pages which would annoy the ecclesiastics.
>
> It was impossible, however, to stop the progress of science. . . .

Lapin and Fridman fill 30 pages of their book, which is 108 pages long, before they mention America's foremost early primatologist, Robert Mearns Yerkes. This takes a bit of doing, since Yerkes was very much on their map as well as ours. But at last, having nodded as it were in his direction, they do keep him in mind, showing anxiously wherever possible that their I. Metchnikoff, the founder of primatology in Russia, was a little ahead of the American all the way along the line.

At this particular moment, when government cutbacks in expenditure have seriously hampered scientific research in the United States primate centers, Russian research workers at Sukhumi and elsewhere are no doubt feeling more relaxed than usual. Unlike America's primatologists, they have recently been given a boost. Dr. Geoffrey Bourne, director of the Yerkes Regional Primate Research Center at Emory University, in Atlanta, Georgia, attended a meeting of primatologists at Sukhumi in 1966–67 and found himself one of five or six compatriots there. As he recalls it, the American party was fairly large in comparison with other foreigners.

"There was only one Englishman," he said, "and I think they had two Czechs, one Rumanian, and two Poles. All the others were Russian, and they were obviously pretty tense because a good deal was at stake for them. A number of important scientists on government committees had come down to attend the meetings, and during the talks the Sukhumi people proposed opening three new primate centers in different places in Russia. Afterwards we heard that the proposal had been approved in Moscow."

Though the book does not insist on this point, Dr. Yerkes appeared on the American scene a good many years before 1925, the year the Tennessee "monkey trial" hit the headlines, but they give the trial plenty of emphasis in an indirect way. The "idealogical atmosphere" in the United States in the 1920's, they remark, wasn't conducive to work that depended on the similarity of man and the other primates. Yet Yerkes was nearly fifty in 1925, and he was, of course, a convinced Darwinian from way back, with lively hopes of setting up a breeding station for primates that could be used in medical and psychological research.

The word "primate" carries the connotation that apes, monkeys, and lemuroids (or prosimians) are at the top of the mammalian heap, with Homo sapiens over all. It is no doubt a boastful claim, but it is not unjustified, since humans do seem able to outguess the other mammals: at least if we can't outguess them, we can exterminate them.

In spite of, or perhaps because of, the Scopes trial, it is safe to say that most Americans today accept the Darwinian theory of evolution or some modification thereof, and take it as fact that we are closely related to the other primates. Certain of them are like us in some ways, and others are similar in quite other ways. The chimpanzee, the gorilla, and the orang-utan look most like us and are closest to us in size, yet the doglike baboon seems to resemble us more closely than any of these in certain of his blood characteristics. Macaque family structure is very like that of some human social groups—and so on. Because of all these likenesses, research students in medicine, surgery, and the other curative sciences, not to mention psychology, usually prefer to work with primates rather than with so-called laboratory animals—guinea pigs, rats, dogs, etc.—but primates are difficult to get, the commonest of them costing more than most other laboratory subjects. Furthermore, they are wild animals, undomesticated—at least up to now—and many species are being squeezed out of existence by proliferating mankind. Many of them are delicate, and most are harder to handle than guinea pigs, being larger, stronger, and far touchier. A laboratory using primates must be fitted with special living quarters for them, with a regulated temperature and strong cages, for it is surprising how quickly a baboon or an orang-utan, for instance, can pick a lock. Besides, because wild-caught primates are infested with parasites inside and out, they must be quarantined after capture and thoroughly cleaned up. Sometimes these precautions are not adequate: recently a number of imported green monkeys were responsible for a serious epidemic among laboratory attendants in Germany—the "Marburg disease," of which seven people died.

So the final cost of such animals comes high. Moreover, the wild cannot go on supplying them at the rate they are needed. In *Science*, December 4, 1970, an article by Charles H. Southwick, M. Rafiq Siddiqi, and M. Farook Siddiqi, *Primate Populations and Biomedical Research*, calls for close attention. The authors first list recent discoveries made in the laboratory through the use of various primate species, and observe that no single species can serve all research interests. Nor can scientists predict which species may prove to be key models in a particular problem. But there is great danger in the "systematic utilization of primates in biomedical research" because of the steadily accelerating disappearance of the world's fauna, of which the primate is especially vulnerable to destructive forces. According to the latest figures released by the International Union for the Conservation of Nature, there are 275 mammal species on the rare and endangered list, and of these 49, or more than 15 percent, are primates. These species are in special jeopardy because of the loss of their habitat, human competition for food, and—sinister sign—the increasing use of primates as food. There is also the demand for primates by both pet dealers and biomedical researchers. More than a hundred thousand primates are used for research each year in the United States alone.

The writers, after recapitulating cogent arguments for conservation and disturbing details about the vanishing primates, add, "We believe there is a danger of undue emotionalism about primate conservation before adequate field data are available. It is likely that biomedical research will receive the brunt of blame for many problems. When shortages of primates occur, the most convenient and visible scapegoat is the research laboratory."

They feel that two major types of program should be initiated as soon as possible: a well-planned program of population research to provide more accurate data on the ecologic status and reproductive biology of important primate species, and *active conservation programs for all endangered species directly utilized in biomedical research.*

In other words, breeding colonies. Only this way seems to offer hope. But breeding chimpanzees, for example, is a slow process: they do not become sexually mature for at least eight years, usually more, and the chimpanzee gestation period is nearly as long as ours.

When the National Institutes of Health inaugurated seven primate centers and started them operating within a short space of time, the action represented an ambitious, imaginative, elaborate program. That the centers are working so well after only ten years reflects a good deal of credit on the project's planners.

Of course, the United States had other primatological institutes before 1960, most important of which was Robert Mearns Yerkes' Anthropoid Experiment Station at Orange Park; and there have long been primate laboratories at some American universities and medical colleges. The air force maintains its own primate observation groups, and so, in a smaller way, does the army. The chief difference between these places and the centers founded and launched by the NIH is one of scope.

Yet in spite of the fact that monkeys exert a fascination over most people—possibly a fascination of repulsion rather than attraction, but a definite one, nonetheless—the centers are not well known to the general public. Those Rotarians of Jacksonville were exceptional in being aware that such a place as the Yerkes Station existed, but after all, it was on their doorstep. Away from the centers few people know anything about them or, indeed, about primatology generally. There has been a lot of publicity about the Salk vaccine and its dependence on the rhesus macaque, but Salk did not work at a NIH research center: there were none in those days.

For a while, I wondered why more notice has not been paid to these places. After all, every newspaper editor knows that a monkey story is always good, even if it is only a routine item about an escaped pet. Then I found the answer in a book by Ada Watterson Yerkes, wife of Dr. Yerkes and co-author with him of the standard work on great apes. In *Yale in Florida, 1929–1939,* describing the early days of Yerkes' venture at the Anthropoid Experiment Station at Orange Park, she wrote, "Non-scientific articles about the chimpanzees occasionally appear and many more would be written if they were not frowned on by the authorities. Notoriety is no help to scientific work, nor curiosity of the newspaper and motion picture type. To the public all simians make an appeal too humorous to be favorable for a serious undertaking."

One knows what she meant, even though things have changed to some extent since those days. I once had an English friend who kept a chimpanzee in London and complained that the animal created a sensation every time he took it out. "How would *you* like it if everyone who looked at you burst out laughing just because you're a chimpanzee?" he demanded indignantly. "No wonder the poor little chap's neurotic."

No doubt, it was because of this practice of secrecy that I first heard of the centers only through a fluke, which started me off on what was to prove a long pilgrimage. An architect I met at a cocktail party told me of a colleague of his who, he said, was working in Texas at a place where two thousand baboons were to be housed.

The gossip startled me. Years ago in the Belgian Congo I kept, or at least chaperoned, a troop of eight or nine baboons, the olive-yellowish species. It was possible to carry on an acquaintance with these animals as I did, free of rancor, only because I was living in the forest with no European neighbors to make objections. I have never forgotten my experience with *Papio anubis*. One baboon alone is a handful: nine were—to put it mildly—distracting, and I could not imagine how two thousand could be maintained in a big city—Houston, I thought the architect had said—without catastrophe for all concerned. Come to think of it, why should anybody keep baboons in Houston? What were they doing in Texas anyway? When I asked the architect this, he said that scientific matters weren't really his line, but he believed that baboons are prone to atherosclerosis and that these animals were being studied by doctors who hoped to find out more about the disease from them. He added that atherosclerosis is one of the more worrying kinds of heart disease.

Draining my Scotch and soda, I brooded about the matter. "How do baboons get it?" I asked at last. "How did those doctors find out they had it in the first place?"

He couldn't tell me. All he knew was that the baboons were being treated so that the researchers could find out what made the atherosclerosis worse. "I think some are fed on a high cholesterol diet," he said, "lots of fatty steak and bacon and eggs and all that. Then they're tested to see if their arteries are hardening or something. And some others are subjected to the noises of civilization, stress and strain, you know—police sirens, fire sirens, racing cars without mufflers, riveting. . . ."

While I was looking for another drink, he disappeared.

For days afterward, I couldn't get those baboons out of my head. I tried, but failed, to imagine Houston streets thronged with them. A baboon is a fairly large beast that looks like a cross between a lion and a dog—a long-muzzled dog with eyes too close together. It usually walks on all fours; though when it wishes to do so, it sits upright and uses its hands like a—well, like a primate. The biggest animal in my Congo troop was especially quick at catching the balls I threw to him, but they would all have made good fielders. Their tails are worn with a jaunty lift near the root. Come to think of it, jaunty is a good word altogether for baboons. There is something attractively crazy about them, and to my mind they are beautiful.

My overheated imagination, after that cocktail party, saw them pacing majestically through the city, being subjected to the stress of urban noises, occasionally dropping in at their doctors' offices to be checked for hypertension. Whenever I met somebody from Houston, I asked eagerly about the colony, but no one seemed to have heard of it. Their ignorance is easy to understand since the baboon colony, as I later discovered, is not in Houston at all but in San Antonio. There, outside the city, they live at the Southwest Foundation, a thousand of them—not two thousand—dwelling pacifically in cages, family groups all together.

The real point of all this is that my search, however misguided, led me to discover many more facts about primates in America. I found that in addition to the baboonery in Texas there are, among others, an experimental breeding station at San Diego Zoo, several colonies (mostly rhesus) living free on islands near San Juan, Puerto Rico, a primate laboratory at Johns Hopkins, another connected with the New York University Hospital, and, most important of all, the seven large Regional Primate Research Centers managed under the auspices of the NIH, the National Institutes of Health which are themselves centered at Bethesda, Maryland.

As I made these discoveries, inevitably I kept coming upon the name of Robert Mearns Yerkes. This did not surprise me, since anyone who has even dabbled as I had with apes and monkeys has heard of Yerkes, but I knew little about him then (though he was alive until 1956), only that he was a Yale professor who kept a lot of chimpanzees somewhere in Florida. The regional primate centers were merely a plan in 1956, but the more I saw of them and heard of Yerkes, the more I realized how closely they follow the pattern he conceived years ago.

Yerkes was born in 1876, the eldest son of a struggling Pennsylvania farmer who thought the boy should stay at home and help support the family by working the farm. But Robert had his heart set on getting an education, and no amount of pressure from the elder Yerkes could deflect him. According to his daughter Roberta, Mrs. Brand Blanshard, he was unusually stubborn, so much so that when his teachers tried to break him of left-handedness, as was the pedagogical custom of the time, he flatly refused to be changed over. He himself wrote that as a child he was "moody, strong-willed and unsuggestible." His father continued trying to persuade him to leave school, with mounting energy as the family financial situation deteriorated, but Robert remained unconvinced: he had almost decided that he wanted to be a doctor.

By dint of outside jobs and help from relatives, he got into Ursinus Academy at Collegeville, Pennsylvania, where a year qualified him to enter Ursinus College in the same town. He lived with an uncle and paid for his keep by doing chores for the household—looking after horses, mowing grass, and cooking breakfast. In the summer he worked in his father's fields.

In 1897, at the age of twenty-one he found himself hesitating between two paths—studying medicine in Philadelphia or doing graduate work in biology, psychology, and philosophy at Harvard. The Harvard prospect tempted him. He had already come to the conclusion that the practice of medicine was not for him, and when a family friend offered to lend him a thousand dollars to help him do as he wished, he was glad to accept. At Harvard he found himself in a hotbed of exciting new ideas. He studied philosophy with Josiah Royce, William James, and George Herbert Palmer, and biology with E. L. Mark and G. H. Parker. He felt attracted to zoology and psychology, where it seemed to him most of the action was, until at last he asked Royce's advice as to his future. The philosopher suggested that he combine his two favorite subjects in a new study called comparative psychology, and work under Hugo Münsterberg if Münsterberg would accept him. Münsterberg did accept him as a graduate student, and Yerkes soon became an instructor in the department, where he remained for the next eighteen years.

An enthusiastic writer of letters and diaries, he later summed up this part of his life in the compendium *Psychology in Autobiography:* "Always my research has been more nearly physiological than psychological, for I have dealt with problems of behavior, not with experience. Therefore my constant use of the descriptive term psychobiology." He said he rejected John B. Watson's behaviorism and preferred to call it objectivism. He had never been able to accept "the extreme objectivism" of certain eminent biologists, for he considered it dangerous in its restrictions and negations. Nevertheless, he said, he was not prejudiced against the study of consciousness, thinking of it as "at once the most fascinating and the most important in biology. . . . That my path is not obviously directed toward this end needs neither explanation nor apology. My course in research is pragmatic."

At that time there was no such thing in America as a laboratory devoted exclusively to the study of animal behavior, and Yerkes was conscious of the need for one. He often wished that there was "an institute of comparative psychology" ready at hand for people like himself. Lacking it, like other workers in his field, he studied crabs, turtles, crawfish, and frogs, all of which were easily obtainable and not much trouble to keep. Then for a time he concentrated on a tiny mammal and, in 1907, published a book—his first—about this creature: *The Dancing Mouse: A Study in Animal Behavior,* describing a workmanlike piece of research which testified to his careful technique. Starting with a pair of these mice, for two years he had observed, tested, and kept notes on them until he had a colony of four hundred. At this point they were carried off all at once by an epidemic. It was a tragedy, of course, but on reading Yerkes' book, one wonders what more he could have found to say about them anyway. Dancing mice are unfamiliar to most of us nowadays, having long since been pushed out of the pet market by Syrian hamsters. They are small even for mice, are usually colored rather like pinto ponies, and are ceaselessly—or almost ceaselessly—active, pausing in their dancing only to eat and sleep. A dancing mouse keeps twitching and turning, or whirling, in one small circle as if chasing its own tail: occasionally, however, it may be joined by another mouse or two, and then, with its companions, it runs in a slightly bigger circle, nose to tail. Japanese mouse fanciers used to keep these pets in elaborate cages fitted out with miniature waterwheels, staircases, and castles to fidget about in.

The problem Yerkes set himself was to find out how the dancing mouse got that way. After patiently watching his specimens, reading what he could find about the species, puzzling, experimenting, and reasoning, he reached the following conclusions: (a) The animal was probably bred deliberately by Chinese fanciers centuries ago, from mutants. Occasionally in a nondancing-mouse family one does get a whirler or dancer. (b) Dancing mice are not invariably either right- or left-handed when they whirl: some go one way, some the other, and some take turns in both directions. (c) Dancing-mouse females are not good mothers. They make poor nests, and when babies wander outside the nest's circumference, the mothers do not carry them back in their mouths, as any self-respecting common house-mouse mother would do. Dancing-mouse babies can die of cold and starvation a mere two or three inches away from home. (d) Adult dancing mice are *deaf,* a condition that may very well explain their lack of dizziness. Some young mice—not when they are born, but during adolescence—do, for five days at most, hear a little before the racial silence descends on them forever.

To design the experiments necessary to prove this last item and then carry them out was a considerable achievement, and Yerkes' reputation waxed accordingly. He published more papers in scientific journals on habit formation, and imitation, and Pavlov, and instinct. He prepared his second book, *Introduction to Psychology*. Meantime, at the back of his mind, he went on thinking of how much an institute of comparative psychobiology would help himself and his colleagues in their work.

In 1910, with an associate named Bloomfield, he published in the *Psychological Bulletin* a paper entitled "Do Kittens Instinctively Kill Mice?" Stirred to action by another behaviorist's contention that cats learn the business of mousing through imitation, Yerkes and Bloomfield brought up two litters of kittens from birth without their mothers, and exposed them to mice—or vice versa—at various intervals. In each litter one kitten definitely reacted and killed a mouse, after which its companions seemed to get the idea and followed its example. The men concluded that kittens by instinct, without opportunity for imitation and without other experiences with felines, etc., do develop predatory behavior. They also decided that instinct becomes increasingly difficult to evoke as the animals age, so that if one wants a good mouser, one's kitten should be exposed to mice while it is still young.

Yerkes continued. He worked with rats in a maze, studied color vision in the ring dove, observed the behavior of crows. In 1915 he returned to mammals with "A Study of the Behavior of the Pig *Sus scrofa* by the Multiple-Choice Method." Through all these studies he was evolving methods that were later to be useful to the government, and later still to general education and industry: he called these disciplines the measurement of mental ability.

In 1910 an ex-student of Yerkes, Gilbert V. Hamilton, set up a small laboratory of his own in Montecito, California, where he kept a few monkeys and an orang-utan and made behavioral studies of them. Yerkes went out to visit him during a sabbatical year, 1914–15, and joined in the studies, which he found fascinating. His "daydream," as he called the nebulous plans he had long entertained, now moved into a new phase when he saw what he wanted and needed—an institute to maintain varieties of primates where he could breed and rear them over generations, observing them and coordinating these efforts with existing biological institutes or departments of research. Such a program was sure to be rewarding to many of the sciences, he reasoned: he could visualize medical, sociological, and psychological points of contact between the lower primates and the human condition. Chimpanzees, for instance, are susceptible to the human cold and other human diseases, and when it comes to psychology, he reflected, habit formation and social relations among many of the primate species are strikingly similar to our own. This last is true of a number of primates, even those so-called lower species on the family tree (if we accept the postulate that the four anthropoid species are the highest)—macaques, baboons, and others—but what fascinated Yerkes most of all were the anthropoids, or so-called great apes: the chimpanzee, the gorilla, the orang-utan, and the gibbon.

In the Western world there was little practical knowledge of these anthropoids. A few gibbons, the smallest and easiest to handle of the four, had for some time lived in American zoos, but as they are the least human-looking of the great apes—many primatologists today dispute their position as anthropoids at all—they caused little controversy between amateur Darwinians and Fundamentalists. For the same reason they did not attract Yerkes very much. With the chimpanzee it was another story:

one could scarcely look at a chimpanzee without thinking of Darwin. There were numbers of them about, too, in circuses and zoos, though in the United States they were difficult to find and expensive. Orang-utans were scarcer, though as we have seen Hamilton had one. The true rarity among anthropoids was the gorilla.

For one thing the species had not even been discovered by white men until early in the nineteenth century—or, if a certain seventeenth-century description does indeed refer to a gorilla, it is too vague to depend on. And mere discovery, when it came, did not mean much to collectors. Usually when a new animal was spotted, the hunters of the time hurried pellmell to collect specimens of it—alive if possible, but if not, dead.

For obvious reasons, few of these men were eager to bring back live gorillas. Those they did bring were too young to be dangerous, and inevitably lived only a short while. Yerkes gives the sad little list of the first gorilla babies to be seen in our Western world: Jenny, or "Wombwell's gorilla," arrived in England in 1855 and died very soon; the Falkenstein gorilla appeared in the Berlin Aquarium in 1876, and died within sixteen months; in 1885 a young male was brought to Paris but survived only a short time.

Dr. William Hornaday of the Bronx Zoo reported in 1904 that only one gorilla had ever arrived alive in the United States, and it lasted a mere few days. Later two more specimens were acquired by the Bronx Zoo—one in 1911 and the other in 1914. The latter lived not quite a year, the former only ten days. In 1915 Dr. Hornaday wrote in the *Zoological Society Bulletin,* "There is not the slightest reason to hope that an adult Gorilla, either male or female, will ever be seen living in a zoological park or garden. . . . It is unfortunate that the ape that, in some respects, stands nearest to man, never can be seen in zoological gardens, but we may as well accept that fact—because we cannot do otherwise."

It took years to disprove this gloomy statement. Not until 1956 was the first gorilla baby successfully born in captivity, in Columbus, Ohio.

Many of Yerkes' colleagues were, like himself, fascinated by the thought of working with primates, and it was tantalizing for him that some of them were able to maintain subjects for observation. In France, the Russian I. Metchnikoff was studying the effect on primates of human diseases, and trying out remedies for these. Then there was Hamilton in California, working on nervous and mental disorders in monkeys and correlating his findings with similar diseases in humans. In 1912 the Prussian Academy of Science brought nine wild chimpanzees to Tenerife, and the zoopsychologist Wolfgang Köhler moved to the island to work with them. Two of the chimps died, but Köhler had the rest—seven chimpanzees—at his disposal until the station closed down in 1918. It was his work above all that made Yerkes' mouth water, for he was finding out fascinating things by means of intelligence tests.

Köhler had discovered that chimpanzees will readily use sticks as tools to lengthen their reach when fruit or other desired food is placed just too far for their arms to stretch. Some of his apes were even more clever and were able to put together two parts of a jointed stick. They knew how to make themselves taller by piling one box on top of another, and would repeat the exercise when necessary until the box tower stood three or four high. Many of their reactions led to the conclusion that they were capable of reasoning. . . . If only he could study such animals! thought Yerkes longingly, and he tried again, as he had tried many times in the past, to persuade Harvard to spare money for such a project. The attempt was fruitless as usual, and he returned to the crows and pigs.

The Yerkes now owned a farmhouse in Franklin, New Hampshire, where the family went every summer and where in the fields near the house Yerkes and his colleagues could work with various animals. He referred to this farm as the Franklin Field Station for the Harvard Laboratory of Animal Psychology.

One day in the early summer of 1915 his attention was caught by a news item from Havana. A chimpanzee had been born there to a captive mother. "The event was notable," he wrote later, "because there is no previous record of the conception and birth of one of the great apes in captivity anywhere in the Western Hemisphere." (Lapin and Fridman say flatly that it was the first event of its kind in captivity anywhere.)

At Quinta Palatino, the estate outside town where the birth had taken place, the chimpanzee's owner was happily receiving congratulations from naturalists all over the world, including Metchnikoff. She was Madam Rosalía Abreu, a lady of great wealth, daughter of a Cuban businessman, who had an overwhelming passion for apes and monkeys. To acquire such animals she spent enormous sums, and once she had them, she kept them in what she considered the best possible conditions.

As might be expected, Lapin and Fridman in their account of Madam Abreu are pretty scathing. "Primarily, monkeys were kept there for the amusement and dilettante observations of the wealthy owner," they wrote. "Anthropoid apes were taught 'good manners;' they were trained to eat at table, use a napkin, spoon and knife, and the like. For the night the animals were placed in special rooms where they were given mattresses and blankets. . . . The owner of the colony was amazed at the 'intelligence' of the anthropoids' behavior. She was convinced that apes were capable of religious feeling."

Instant psychologists might leap to the conclusion that Madam had a frustrated maternal complex. In fact, she had four or five children of her own. Presumably she just happened to like apes and monkeys, that's all. She and her husband were divorced, and the Yerkes record never mentions Señor Abreu. Life at Quinta Palatino, her estate, was quiet, but the imminent birth of that chimp baby was important, and she had invited Dr. Louis Montané, an anthropologist living in Havana, to be present at the lying-in. The following autumn he read a paper about the occasion to the Cuban Society of Natural History, and from this paper we learn that the infant ape was baptized.

Robert Yerkes wrote to Madam Abreu as soon as he had her name and address, in July 1915. (Their correspondence is to be found in the Yerkes Papers in the Medical College Library at Yale.)

"I am deeply interested in the study of the habits, instincts, and intelligence of the anthropoid apes (especially the orang utan and the chimpanzee) and knowing that you have kept a number of them successfully on your estate in Havana I am taking the liberty . . ."

He asked many questions about the little chimpanzee, and was gratified when Madam Abreu quickly sent back full replies, as well as a photograph of the infant. She wrote happily, "The first time he reconize me was the 6th of Mai, being only nine days, when I called him by his name, Anumá, in souvenir of the monkey god of India, he look at me and scream hu! hu! The mother is now making him walk by holding him by the arms, and some times put him on her back, like horseback, and like the negroes do in Africa with their babys."

Copious correspondence followed, with Madam answering all Yerkes' later inquiries with equal zeal, though she was busy coping with other interested scientists as well. She seems to have recognized in Professor Yerkes a soul mate. More than once she invited him to Cuba where he could look at her animals in person, but as he explained, he didn't have the money for such a trip and the department would not pay his way. In the spring of 1916 she offered to help in practical ways if he should wish to start a primate laboratory there in Havana: she said she would supply land and animals as long as the primates "would not be submitted to experimental medical purposes, which would cause them suffering."

"Come and see my Cuban baby chimpanzee," she wrote in December, "which is the sweetest creature . . . that ever I have seen."

That Yerkes *was* trying hard to do something about Quinta Palatino is indicated by a technical article he published in *Science* that same year—

1916—"Provision for the Study of Monkeys and Apes." Lapin and Fridman, referring to this article, hastily footnote the reference:

> The first pronouncement in favour of the organization of monkey breeding stations for experiments was made by I. Mechnikov [Metchnikoff]. On May 15, 1915, at the celebration of his 70th birthday, Mechnikov said in his answering speech, "Experiments should be carried out at monkey stations where attempts should be made to breed monkeys.

Ah, yes; but Yerkes' plans went much further than that. His article was a general outline of the daydream-become-program as it now arranged itself in his mind, and is remarkable chiefly for two things: the scope of studies he was confident could be encompassed in such a colony, and his modest estimate of what it would cost. He thought that six assistants would suffice, of whom one at least should be trained in comparative physiology. An original endowment of a million dollars would yield enough, he figured, for the annual budget of fifty thousand dollars. Perhaps it is as well that these optimistically low estimates were never put to the test. As matters turned out, Yerkes' plans—like Madam's and the rest of the world's—were abruptly cancelled when America entered the First World War. As he wrote in his book *Almost Human,* for several years afterward he heard only indirectly and rarely about the Cuban baby primate.

2

Orange Park

Yerkes' contribution to the war effort was unconventional but demanding. The psychologist who had spent so many years investigating the intelligence of animals now found himself immersed in the study of man's intelligence, and he had little time to wonder how things were going at Quinta Palatino.

A paper by Daniel J. Kelves in *The Journal of American History* for December 1968 tells of this venture. Even among psychologists, the writer says, intelligence tests commanded little more than shaky respectability until 1916 when the Binet test was revised by the American Lewis M. Terman. Among laymen this shakiness persisted long beyond that date, but as soon as the United States entered the war Yerkes convened the Council of the American Psychological Association to discuss the military uses of psychology. At the meeting he was appointed chairman of the Committee on Psychological Examination of Recruits.

Originally this committee proposed only to test recruits suspected of mental incompetence, but soon, encouraged by their chairman, they decided to go further and classify men so that they might be properly placed in the military service. It was a proposal that met with considerable resistance from most members of the military establishment, where officers were promoted according to seniority, and assignments of noncommissioned men were made on the basis of firsthand personal knowledge, twice a year. The army had no general method for identifying men with special skills.

Yerkes was also chairman of the Psychology Committee of the National Research Council, created in 1916 by the National Academy of Sciences. Fortunately for his cause, another member of the committee, the Surgeon General of the United States Army, kept his mind open and thought there might be something in the new ideas. Through his recommendation, Yerkes was commissioned major and put in charge of a project, provisionally approved, of trying out tests devised by the committee in four cantonments for a start.

The War College's Committee of Training was more skeptical than the surgeon general, but after some months it decided to approve the tests with one qualification—that intelligence ratings alone should not determine a man's selection for promotion: they must be supplemented by a knowledge of personality, appearance, energy, and resourcefulness. Even though the qualification was adopted, after the war the program was dropped. Altogether it was a frustrating experience for the psychologists.

Yerkes was deeply disappointed, but in Kelves' opinion the work was further reaching than he realized. "The psychologists did not leave the military unchanged," he comments. "Recruiting officers were now administering intelligence tests to men of uncertain literacy. In addition, Scott and Yerkes had laid the foundations for a scientific personnel system. . . . Pockets of appreciation for the tests remained within the regular military, particularly among the younger generation of career officers. . . . Even more important, the wide use of the examinations during the war had dramatized intelligence testing and made the practice respectable. Gone were the public's prewar wariness and ignorance of measuring intelligence."

When the war ended, Yerkes remained in Washington as chairman of the Research Information Service of the National Research Council, but he missed his research studies, and he had not forgotten the plan for a primate laboratory.

In 1923 he got his break, unexpectedly. It was summer, and he had moved to the farm with the family as usual, when he received a letter from William Hornaday telling him that the Bronx Zoo was boarding a pair of little chimpanzees temporarily for their owner, and that both the male and the female were for sale. The situation was nothing out of the ordinary: he would not have bothered to write about it, said Hornaday, but the male chimp was most unusual in appearance and behavior: he thought it might even belong to a new species. At any rate he felt that Yerkes should have it for study.

"His letter reawakened my long-suppressed desire to study the behavior of the great apes," Yerkes said later. He went to New York, and as soon as he laid eyes on the chimpanzees, he lost his heart. The female he considered ordinary enough—in fact, less than ordinary, being "in wretched condition"—but the male was really a rare specimen. Yerkes had never seen another animal to match him. In fact, though he was never to know it for a certainty, the chimp was indeed something special: he was a pygmy chimpanzee, or Bonobo. Today many experts consider the Bonobo not a mere subspecies of chimpanzee but a separate genus in its own right.

One snag about acquiring him was that the purchaser had to take both or neither, and Yerkes did not want the female. A bigger difficulty was that the owner was asking the huge sum of three thousand dollars for the pair, though the going rate for chimps was from two to five hundred dollars apiece. But now that he had seen the little male, Yerkes simply had to have it.

"All of my savings, apart from life insurance, amounted to less than $3,000," he said in his reminiscences. "The conflict between desire and discretion was acute. Ada and the children supported my enthusiasm and venturesomeness, or I could not have made up my mind to offer Mr. Lewis $2,000 for the pair. The offer was promptly accepted."

He named the male Chim and the female Panzee. Back at the farm, they "took to our New Hampshire woods and pastures like ducks to water. . . . Chim proved a treasure. He was physically perfect, healthy as could be, quick to learn, and he gave varied indications of exceptional intelligence and temperament. Panzee was a perpetual headache and drain on our sympathies . . . ," but she, too, improved in health and spirits in New Hampshire's sunlight and air. "We were therefore hopeful of restoring her to perfect health," writes Yerkes in *Chimpanzee Intelligence and Its Vocal Expressions,* but in the autumn, when the family, chimps and all, moved back to Washington, D.C., she began to lose ground, and in January, in Havana where Yerkes had taken her, she died. The autopsy showed that she was tubercular. Chim had escaped the infection: his reaction to the intracutaneous tuberculin test, which Yerkes arranged for soon afterward, was completely negative.

Roberta Yerkes Blanshard remembers both animals very well, and admits that it was hard to love Panzee in her sick and apathetic state. Still vivid in Mrs. Blanshard's mind is her father's enthusiasm when Chim built his first nest. This excitement is easy to understand. It was the first time in the history of the world that an anthropoid's nest had been made in New England; very likely it was the first time one was built in a birch tree anywhere.

To make a nest, a chimpanzee sits in a bush or a tree and pulls down the branches growing near it. Then it stamps them down until they are beaten into submission, and the animal is on a sort of platform with a depression in the middle. "To see the birches of a New Hampshire hill pasture filled with chimpanzee nests makes one feel queer," wrote Yerkes in *Chimpanzee Intelligence*.

The move to Washington gave young Miss Yerkes yet another vivid memory of Chim and Panzee, for her choice third-floor bedroom with sleeping porch attached was taken away from her for the chimps, and she was relegated to a second-floor alcove. The bedroom and sleeping porch were remodelled into indoor and outdoor cages, between which the apes moved at will. One peculiarity of the sleeping-porch-turned-cage was that its inner-walled area included a window looking into a bathroom. An occupant of the bathroom could look out on the chimps if the chimps were there, and what was perhaps more important, the chimps could peer in and watch whoever was taking a bath. The apes invariably did watch, too, unless the bather drew the curtains against them; and Roberta Blanshard says that guests sometimes complained about the arrangement, but it was never changed. Yerkes himself looked after the chimps, preparing their food, and so on. Sometimes when a caller or a delivery boy rang the bell, he answered the door in his white laboratory overalls, a chimp on each arm. He enjoyed observing the newcomer's reactions to this sight.

As soon as the apes were settled in Washington, Yerkes wrote to Madam Abreu, breaking a seven years' silence. How was she after all this time, he asked, and did she still have her anthropoid apes? Was the baby thriving? By return mail Madam replied: Yes, indeed, she still had the collection. Anumá was beautiful and healthy, and there was another chimp baby at Quinta Palatino, nine months old. Yet another had been born between these two, but that one, with its mother, had died of pneumonia contracted while travelling (with Madam, naturally) in Europe. She had a dozen chimps of all ages, all healthy. Jimmy, father of the three babies, was now twenty-five years old—an item especially interesting to Yerkes, for the longevity records of chimps in captivity were few and suspect. Indeed, not very much is known even today about the chimpanzee life-span, either in captivity or in the wild.

Madam Abreu's letter was long: clearly she was happy to be able once again to discuss her hobby with a sympathetic correspondent. She had seventy-five animals in all, she said, representing twenty-five species, among which were three orang-utans, two lion monkeys (lion-tailed macaques), a small mandrill, a colobus, twelve "golden yellow" baboons, one chacma, a pair of "woolen" (woolly) monkeys "the most affectionate you ever saw," and "a lovely gibbon, who sings when she sees me appear in the garden. . . . I am specting more from Africa, in the spring perhaps a gorilla. I had bought one that arrive half dead in New-York."

There seems to be no record in the Yerkes-Abreu correspondence of the proposition Russian scientists had been making to her, requesting her cooperation in an experiment that proves to be our old friend crossbreeding. It is left for Lapin and Fridman to mention it:

> In the 20's the prominent Russian biologist I. Ivanov made plans for carrying out experiments at the station [i.e., Quinta Palatino] on the hybridization of anthropoids and cross-breeding the latter with man. Madam Abreu assured Ivanov and the management of the Pasteur Institute in Paris of the possibility of staging such experiments and even announced her intention of buying some more chimpanzees for the purpose. But the trip to Cuba did not take place: according to the archives, a cable was received from Madam Abreu a few days before Ivanov's departure informing him of the death of the male chimpanzee and orang-utans necessary for the experiment, and of her apprehensions of being discredited in the eyes of the people of her set. The lady owner of the colony had evidently received a letter, like Ivanov himself, from the Ku-Klux-Klan, containing threats and curses for the intention of carrying out experiments "abominable to the Creator."

Ivanov or no Ivanov, Madam's letter to Yerkes made him restless in the old way. In the following months they exchanged more letters and many cables. Animal dealers in America frequently offered Madam primates newly arrived from abroad, and she needed expert advice on whether or not to buy from someone on the spot. Yerkes was able to inspect some of the animals and give her his opinion of them, and she appreciated his help. In the past dealers had often sold her sick or moribund primates, she said, and their prices were growing exorbitant. Yerkes replied that though he agreed with her as to prices, she must keep in mind that the market *was* difficult. Chimps were hard to get at any price, because "the showmen" were snapping them up, especially those people who trained animals for the moving-picture industry.

Finally he asked Madam a serious question: Had she thought any more of establishing a research station in Cuba for the study of primates? Such a station could be affiliated with Yale or Harvard, and studies of behavior and genetics could be made with her animals. If she was still interested in the idea, he said, he would come to Havana to talk it over, and when Madam assured him that she was, he made a preliminary visit to Quinta Palatino in January 1924, and the pen pals met at last. It was during this visit that Panzee died.

Though he was restless at the National Research Council and eager to leave, Yerkes' sojourn in Washington did him good, because it improved his prospects of creating a primate laboratory through the contacts he made there: he learned where to ask for money. In the end a corporation underwrote his two trips to Cuba and the summer program which he planned with Madam. As a result, a party comprising himself, two assistants, and a secretary went to Havana in July of that same year, taking Chim with them.

Working on Madam's estate among her ape cages was bound to be ticklish at times, no matter how cooperative she and her keepers tried to be. It was a situation loaded with potential trouble. After all, the animals were Madam's; the expedition was there on Madam's sufferance, and all the experiments had to meet with her approval. She was intelligent and quick to understand most of Yerkes' explanations, but she *was* in the way. In *Almost Human* Yerkes is scrupulously careful not to complain, and certainly a good deal of useful work was done in Cuba for the cause of behavioral psychology, but much more might have been accomplished if only. . . .

However, that was the situation, and it could not be altered. Yerkes finally arranged his work so that a good deal of it was done directly through his hostess, who willingly answered question after question about her apes' behavior from years of experience. If sometimes Yerkes did not agree with her interpretations, if they seemed to him impossibly anthropomorphic, he did not say so, and his book is very tactful. With photographic illustrations and text, it gives a good idea of how the apes lived.

Some of the estate was taken up by conventional landscaping, with flower beds and ornamental pools according to Cuban custom. Palms, fruit trees, and tall shrubs were in profusion everywhere, but they grew especially thickly in the part of the grounds reserved for the primates. A half-circle of cages stood close to the house, fifty to a hundred feet away: they varied in size from small enclosures for marmosets—three or four feet across and four to five feet high—to the much larger cages that housed the anthropoid apes.

Jimmy, the *pater familias* chimp, with his mate and infant occupied a cage nearly sixty feet long by thirty by twenty. Each great-ape cage was provided with boxes to serve as nests or beds, and some of the cages—though not all—were divided, with the nest box or boxes constructed in a kind of bedroom. Yerkes approved heartily of bedrooms for chimpanzees, kept separate from what he called the living room; he only regretted that all the cages were not arranged in the same way. This was because every night, he explained, Madam took the orang-utans and most of the chimps indoors with her, where they slept in smaller cages in or near her bedroom.

"Her idea is that they are in this way protected from risk of taking cold or contracting pneumonia from exposure during the cold nights which come not only in winter but occasionally at other seasons," he wrote in *Almost Human*. "To such animals as happen to be in the house, she regularly gives milk at eight o'clock or when she returns home if she has been away for the evening. Those animals which are not left in their cages overnight are taken into the house from four to seven in the evening, in accordance with the season and the condition of the weather. . . . In the morning between six and seven o'clock they are returned to their outside cages. Jimmy alone, of the chimpanzees, is never removed from his outdoor cage," because, Madam believed, he was too big and bad-tempered to handle. Yet other animals, such as Anumá, who were also big and strong, were moved.

Yerkes thought the whole business unnecessary. Jimmy stayed healthy out of doors, so why should any of them go indoors? "We are inclined to believe that their owner humors the animals somewhat unnecessarily, at the same time indulging her own sympathetic interest in them and her eagerness to be good to them." For Robert Yerkes at Quinta Palatino, those were harsh words.

Yet in another way he considered that his hostess did not err from too much kindness: on the contrary, he thought her somewhat remiss. ". . . there should be a dining-room," he said in *Almost Human,* "to which the animals may be admitted for their meals and where they should behave like well-trained children, eating what is provided and returning to the outdoor cage [i.e., the living room] when the meal is finished." Admittedly, such an arrangement would call for more assistant keepers and more intensive training. He could understand that she might be reluctant to incur such extra expense, but it seemed a pity. ". . . instead of having them eat in a special room or cage she has each individual or group fed in its living-cage. In so doing she misses an important opportunity to vary the routine of the animals' existence and to afford them a chance for new adaptations and new adventures. . . . we should suggest a special dining-room or dining-cage within easy reach of the other cages, with a long table and chairs and with facilities for use as a playroom or school-room as well as a dining-room."

Yerkes realized that in his position he had no right to complain, but he must often have chafed. He was in sight of the Promised Land, but could not quite get there. Though he was surrounded by this wonderful crowd of apes and monkeys, he could not study them as his discipline demanded, since the deity presiding over them did not understand his methods and aims. Still, Madam was very kind, and Yerkes told himself that he was lucky to have her as a friend. He and his assistants filled many notebooks. It would have been a good summer if he had not met with tragedy within a month after he arrived. Chim died.

The little chimp had been well, even bouncing, at the beginning. He lived in a large cage at the Quinta, and joined in the behavior tests with his usual verve and intelligence, much impressing the working party. Professor Harold C. Bingham, one of Yerkes' assistants, remarked to Chim's proud owner that the little ape "seemed to be interested in the causes or conditions of things, whereas his companions did not." But there was a "respiratory epidemic" in the colony at the time of Chim's arrival, and he soon caught it and then developed pneumonia. He grew feverish, probably even delirious. At least Yerkes got the distinct impression, as he anxiously nursed and tended the little ape, that Chim was actually expecting a caller, some mysterious stranger who was never to arrive.

"Often as I entered the sick-room, I noticed that he was watching eagerly," he wrote, "and although he welcomed me and sometimes even came to me from his bed, he seemed still to be expecting some one. Every footstep in the hall attracted his attention. . . . On one occasion it seemed when I approached him as though he were either mentally disturbed or failed to recognize me."

There were other developments that must have been very painful to Yerkes: "As I sat by him one day, toward the end of the first week, he talked almost continuously, as though trying to tell me something. This unusual performance was very impressive, for it made one feel that, like a person, he was trying to convey certain meaning for which his vocal expression was inadequate. He was not excited, but seemed calm and intent on his task."

There can be no doubt that this experience deeply affected Yerkes. Writing of the ape's death, which occurred soon afterward, he said, "Doubtless there are geniuses even among the anthropoid apes. Prince Chim seems to have been an intellectual genius."

In the end it was Yale, not Harvard, that helped Yerkes to attain his laboratory. Before James Rowland Angell became president of Yale, Yerkes had often talked with him of the dream laboratory and of a special department for advanced students that should, in Yerkes' opinion, be connected with it. As a result, one of Angell's first decisions after he was appointed head of Yale was to create a department according to Yerkes' outline, and Yerkes was invited to occupy a Yale chair in "psychobiology," with the added inducement that he could spend as much time as he chose on research. He jumped at the chance.

In 1925, when these arrangements were made, conditions in the primate market were much as they had been for years. Chimpanzees were available though highly priced; orang-utans were harder to find; but the situation concerning gorillas had somewhat altered.

In 1918 an Englishwoman named Alyse Cunningham obtained a young lowland gorilla and kept it successfully—in London, of all places—for three years, until the animal, John Daniel, outgrew her power to control him. A photograph of John Daniel posing with Miss Cunningham fills one with awe and admiration of both: the lady looks almost as strapping a creature as her ape. But she may have had doubts of her own strength, for she sold him in 1921 to the American circus people, Ringling Brothers: Robert Yerkes heard that they paid fifty thousand dollars for him. Unfortunately, the gorilla died soon after arriving in New York, and to Madam Abreu, who had bid unsuccessfully for him, Yerkes wrote, "It is too bad you did not get John Daniel, the little gorilla from London. Doubtless if he had been sent to you he would still be alive."

In 1923 Miss Cunningham acquired another young gorilla named Sultan or—as he was to be billed in America—John Daniel II. A year later she brought him to the United States, where she exhibited him personally, under the patronage of Ringling Brothers. Sultan actually survived show business for three years, travelling all over America, England, and the Continent: he did not die until 1927, which was the record until that time. Everything considered, Yerkes never thought he would get the chance to study a representative of this rare, delicate species, but he did not waste time longing for it—he was quite busy enough with chimpanzees.

The latest significant contribution to his kind of scientific literature came in 1925, from Russia, from Mrs. Nadazhda Ladygina-Kots, who was married to Alexander E. Kots, curator of the Darwinian Museum in Moscow. Mrs. Kots had taken into her household a little chimp named Joni (or Ioni), and was bringing him up with her small son. She was a worker after Yerkes' own heart. Though limited, as she complained, by the fact that Russia was cut off from foreign scientific aid, she made the most of Joni, keeping copious notes of his general behavior, but concentrating on finding out about his color vision.

Such refinements as the Bradley series of neutral stimuli and color samples of standardized intensity were unobtainable in Moscow, so she constructed her own samples and cards. With these she first demonstrated that Joni could distinguish black from white. Naming "black" or "white" to him seemed to be no help at all—he depended on his eyes, not his ears, to distinguish the colors. (The animal's restive temperament, she said feelingly, often interfered with the progress of the experiment.) Then she tried the chromatic colors, showing a model sample to Joni, who was supposed to choose the matching sample from a collection of many hues. Whenever he got the color right he was rewarded with a tidbit—the "reinforcement" of animal behaviorists all over the world.

The results of these tests were markedly successful. Mrs. Kots discovered that it was the actual color itself, not the brightness of the hue, that was important to Joni when he made his selection. Her conclusions tallied with those reached by Wolfgang Köhler, who tackled the problem from a different angle: both researchers agreed that the chimpanzee's color vision is similar to ours.

Photographs of Mrs. Kots and Joni show an idiosyncrasy in the Russian ape's appearance. Mrs. Kots was a remarkably pretty young woman, but Joni, with his grotesquely long upper lip, looks like the caricature of a chimp.

The Russians, Lapin and Fridman, again take a slightly unusual view of the history of America's primate centers when they speak of the opening of Yerkes' primate laboratory.

> The history of monkey breeding stations and primatological centres in the United States is peculiar. In the 20's, when the use of primates in science became quite necessary, the idealogical atmosphere in the USA was extremely unfavourable for such work, although the country's immense material assets were ample for coping with the problem. In March 1921, a bill was introduced in the State of Tennessee prohibiting the teaching of evolution at educational establishments. The governor approved the bill. Soon ten more states excluded the evolution theory from curricula and then other states followed suit. In 1925, the year it was decided to set up a primate station in Sukhumi, the notorious evolution or "monkey" trial was held in Dayton, Tennessee, the accused, John Scopes, being fined for teaching the Darwinian theory at school. A similar trial was held in tsarist Bulgaria in 1932. In this situation, which was far from favourable for the organization of primate laboratories in the USA, it was not easy to speak about experiments on monkeys from the position of their kinship with man.

Somehow, though, Yerkes managed, in spite of the unfavorable situation, to get started. Two donors—Yale and the Laura Spelman Rockefeller Memorial Fund—pledged enough money to carry Yerkes and his associates through what he thought of as "the second phase" in developing a primate laboratory. He had worked it out, and saw the project as comprising three intimately related divisions: the first for breeding, rearing, and general observation of one or more types of primate; the second—located "by choice" in some large university research center—for the use of primate subjects in investigations; and the third for organizing and carrying out fieldwork, so that the primate types represented in the breeding station could be studied in their natural habitat. "The project still appeared extremely ambitious and exacting but it certainly did not impress competent judges as grandiose," he wrote. Perhaps he was still thinking in terms of fifty thousand dollars a year.

The group decided to specialize in chimpanzees—at least, at first—and Yerkes was busy collecting specimens when, out of the blue, he was offered the chance to study a female mountain gorilla. Congo, as she was called, had been brought to America by her captor and boarded with his relatives in Jacksonville. The Florida climate seemed to suit her, and after the indefatigable Ringlings bought an interest in her, she continued to spend her winters there, as all the circus did. It was Yerkes who did the necessary travelling. Every winter for the three following years, until she died, he spent six to eight weeks in Sarasota studying the gorilla's behavior and intelligence. The contrast of her behavior with that of chimpanzees was fascinating, he reported: her reactions to Köhler-type tests, with sticks and boxes supplied so that she might use them as tools or mounting blocks, were much slower than those of any chimpanzee he had ever tested. Yet after the ten months that intervened between his first and second visits, he observed that she had developed mentally, by herself, in relation to the problems. She had not touched the apparatus during his absence, yet she now had a far better idea of what to do with it. It was a thousand pities, he reflected as he returned to New Haven, that he couldn't work with more gorillas.

Back in New England, after his first gorilla session, he gave some anxious thought to an important question: How would the climate of New Haven affect his animals? "My recent bad luck with Chim and Panzee was difficult to ignore and impossible to forget," said poor Yerkes; but the conditions were not analogous. Chim had picked up an infection in Cuba, and Panzee had already been ailing when he bought her. At any rate, the new chimps would have to take their chances.

Yerkes had already bought a five-year-old pair of chimpanzees from Frank Buck, the "Bring-'em-back-alive" man. The male was named Bill after William Jennings Bryan, and the female was called Darwina—Dwina for short. As Yerkes explained, rather apologetically, in *Creating a Chimpanzee Colony,* "Ordinarily naming an experimental animal seems rather sentimental and less convenient than giving it a number, but in practice we have found it natural and desirable to name our apes. They are highly individualized and their personalities become as distinctive to one who works with them day after day for months or years as do those of one's human companions." In fact, a rule of nomenclature was adopted for the station later on: the females' names were always to end in vowels, the males' in consonants.

Late in the summer of 1926 Bill and Dwina were joined at the Franklin farm by two more small chimps, Pan and Wendy. "Our New England pasture was unique," said their owner, "with chimpanzees swinging from its birches or watching with timid curiosity the herd of grazing cows." With four animals he was able to put into practice some of his earlier ideas: the chimps ate their meals all together, at their own table.

In New Haven, in the autumn, the apes were lodged and studied in a building known to the townspeople as the Manson barn—a red-brick place that had been adapted to this new use. There the primate laboratory was located for its first five years, though the barn was too near the street to suit Yerkes. He complained: "The outdoor cage commonly had a gallery of high school or college boys. Bill early learned to express his attitude toward them by spitting, filling his mouth with water and squirting it at them, or throwing sand in their faces." Roberta Yerkes Blanshard recalls that passing schoolboys were blamed for teaching Bill to spit, but the animal found it especially easy in any case because he had a tooth missing in just the right place. After the laboratory was moved to one of the Medical College buildings the Manson barn, readapted, became a girls' private school, and later still it housed the Yale Department of Astronomy. It is unlikely that anybody thought, as he studied the stars, of the chimpanzees who had once lived there, and the various professors who have in their turn inherited the Manson barn for their offices do not look at all self-conscious as they go in and out. However, one feature at least of the early days has been perpetuated—Bill the spitting chimpanzee turned out to be the first of a long line of spitters. He carried his unlovely skill to Florida when the next phase went into action and he was sent south with other New Haven chimpanzees to breed. There the tradition took root afresh. The technique was handed on, like Freedom slowly broadening down from precedent to precedent, until it moved with the chimpanzees to the new Yerkes Primate Center at Emory University. Today spitting flourishes at Emory, and the staff has become expert at dodging.

The same, unfortunately, can not be said for Russia's Pavlov. Back in the old days when the chimpanzees still lived at Yale, Pavlov attended a World Zoological Congress at the university, and the authorities, naturally wishing to entertain him, took him on a tour of the laboratories. He spoke no English, but he was accompanied by Alexander Petrunkevitch,

Yale's well-known professor of zoology who acted as his interpreter. They entered the room where the apes were kept, while a number of lesser fry —assistant teachers and aspiring biologists—waited reverently outside. One of them, who has since become a professor, well remembers what happened next, and he told me about it recently.

"Suddenly there was a sort of shout inside there," he said, "and the two Russians came out with the laboratory assistant, all talking at once. Pavlov was furious. Petrunkevitch called out to us, 'Oh, the apes bespat him! The apes bespat him!' Pavlov seemed to want Petrunkevitch to say something, and Petrunkevitch obviously didn't want to, and protested, but finally Pavlov had his way, and Petrunkevitch said to us, 'He says this is not science: it is *nonsense.*' "

Four years after it was begun, the general headquarters laboratory—to give the refurbished barn its proper title—had proved to everyone's satisfaction that it was feasible to rear and keep healthy chimpanzees in New Haven. Very interesting work had been done with them, as Yerkes and his wife were able to report. For example, in a food-box test the animals were presented with four boxes of different shapes and colors, one of which held a reward. The chimps were supposed to remember from trial to trial which box held the food, and often they did remember, even when a period of as much as three hours intervened between the first and second presentation. Sometimes changes were rung on the boxes—that is, the food would be placed in a different box of another shape, though the location of the new treasure box corresponded to that of the earlier test. It was found that the chimpanzees persevered in direct proportion to the difficulty of the situation, but if an animal repeatedly failed to find the food, "his assurance would change" (to use the writers' language in *The Great Apes* by Dr. and Mrs. Yerkes) to "hesitation, comparison, watchfulness, and, occasionally, vacillation." His face and behavior might show disappointment, incredulity, mystification, resentment, anger, or depression. Sometimes he lost patience and refused to choose at all. The chimpanzee, Yerkes concluded, is capable to a degree of limited, delayed response hitherto known only in man.

But though they remained well and seemed likely to mature, Yerkes doubted if the animals would readily breed in Connecticut, and he was not inclined to wait in order to find out. He wanted to move on right away to the next development—to found a breeding station in some kinder climate—and such a move would entail the extra expense of considerable building. Yale sent a report on the past four years of work and accomplishment in the laboratory to the trustees of the Laura Spelman

Rockefeller Fund; after studying it, these gentlemen agreed to extend their grant, not only to cover ten years more but to pay the additional outlay.

Now it seemed reasonable to look even beyond the breeding colony, Yerkes thought, to the third and last stage of his original plan—the expeditions for fieldwork, on which people might study anthropoid-ape behavior in the wild. It should be the African wild, he mused. He did not rule out Asia completely, but chimpanzees and gorillas are African.

In the meantime, he had to decide where to put the breeding station. For happy weeks he studied the world atlas looking for a place, while his fieldworkers waited impatiently. He and his advisers considered French Guinea, where since 1922 the Pasteur Institute of Paris had been preparing and maintaining a primate station called Pastoria. They thought briefly of Leningrad, too, because Pavlov was there. But why, Yerkes asked himself at last, need they depend on other nations? The Western Hemisphere would be closer and thus better for their purposes. With his assistants he discussed Central and South America, the various Caribbean islands, and Barro Colorado Island in Panama. What about Bermuda, or Jamaica, or Cuba? There were, however, obvious disadvantages in keeping a lot of valuable animals at even these smaller distances from the home base of Yale, and in the end, in 1929, they bought two hundred acres of land at Orange Park, Florida, fifteen miles outside of Jacksonville. Construction of the buildings began in January 1930, and the initial unit of the Anthropoid Experiment Station—the Yerkes place was to have several names during its Florida career—was completed in June of that year: a laboratory complete with offices and experiment rooms, a service building, and a quarters building with eight animal rooms.

Bill, Dwina, Pan, and Wendy were brought down to Orange Park immediately, and Yerkes with his wife and associates moved in. Soon the four apes were joined by ten more from Pastoria—a truly magnificent gift from the Pasteur Institute, whose authorities sent sixteen animals in all, six of them going to New Haven. In September Dwina gave birth to Orange Park's first anthropoid infant, Alpha, whose father was Pan. A shade was cast over this happy event when the mother died of childbed fever, but Yerkes' judgment in choosing his site was vindicated when, during the following years, many more of his chimpanzees bred and gave birth safely to healthy infants.

In November 1930, Madam Abreu died, and her children, who had not inherited their mother's passion for primates, notified Yerkes in 1931 that as they were settling the estate and giving up the collection, he could have his pick of the animals. He hurried to Cuba and selected fifteen chimpanzees of all ages, from young infants to an adult twenty-five years old. He felt that the older ones might prove to be the most valuable for his breeding colony, since it takes chimpanzees such a long time to mature. Whether it was this infusion of older blood or merely the salubrious climate of Florida, the records of 1940 give cheerful news of Orange Park's success in what Yerkes had set out to do. Where a decade earlier he had owned thirty-five chimps, there were now forty-six—thirty-nine in residence and seven in New Haven.

In comparison, Pastoria's early records make sad reading. If we are to believe Lapin and Fridman, this tract of land near the town of Kindia, in the very middle of chimpanzee country, had a miserable beginning.

> In the winter of 1926, according to I. Ivanov's account, the population was about 30 chimpanzees. Of this number, only one female had lived about two years at the station, while other animals had been there not longer than five months. Ivanov noted the high death rate in Pastoria. A female chimpanzee brought to the station on February 3, 1926, was listed as No. 370. Consequently over 340 monkeys had died in Pastoria. . . . The work carried out at Pastoria was in the line of developing methods of BCG vaccination, preparing some other vaccines and sera and staging experiments in oncology. The basic studies were devoted to problems of tropical medicine.

By 1926, the authors added, not one case of conception had been noted among the chimpanzees or even the lower primates, the baboons, which will usually breed under more difficult conditions. In contrast with Pastoria, at Orange Park chimpanzees bred regularly, and some of Yerkes' apes have lived longer than it was supposed these animals could exist. In fact, at the moment of writing, one of his original animals, Wendy, is still alive at the age of forty-seven.

Having given Yerkes his due, in a manner of speaking, Lapin and Fridman conscientiously listed all of America's primate centers and colonies, then undid their work in one smashing paragraph:

> Thus, there are about 20 monkey breeding stations the world over. Most of them, i.e. some 18 breeding stations, are situated in the United States where until recently, such establishments were all but absent because of religious prejudice and the ban on Darwinism. The Sukhumi Monkey Breeding Station was therefore until recently actually the only big medical primatological centre in the world, situated outside the range of the monkey's habitat. . . .

We are not really concerned with Sukhumi, the colony in Abkhazia in Russia's Georgia, on the coast of the Black Sea, but there are certain points in the narrative of its founding that are worth noting. Lapin and Fridman say what they have to say delicately:

> The immediate impetus to the establishment of the Sukhumi station was given, strange as it may seem, by scientific trends which had nothing to do with its work after it was opened.

What they mean is Dr. S. Voronov. Anybody alive in the twenties who read the Sunday papers in England, France, or America knew about Dr. Voronov and his monkey-gland operation, which was supposed to rejuvenate tired old men. Dr. Voronov transplanted "ductless glands," he said, from primate to man, and that did the trick. He became so well-known that the Russian Institute of Experimental Endocrinology, which seems to have been languishing, reorganized itself and in 1925 got things moving toward establishing a Russian monkey colony so that more glands might always be ready for use.

> Although monkeys have always been outlandish and exotic animals in Russia, it being hard for them to survive in the comparatively rigorous climate, Russian scientists have shown great interest in them. This can be attributed to the fact that in Russia, Darwinism, on which the contemporary scientific use of primates is based, found particularly beneficial soil and gained wide support among scientists.

Yet in the United States, where monkeys are equally outlandish and exotic and where, as every Russian knows, we had the Scopes trial in 1925, Yerkes got ahead of them. . . . Never mind these comparisons: *I* didn't start it. Let's go on with the story, for we are on the verge of something that might explain the Rotarians of Jacksonville and their questions:

> Another factor which led to the organization of a monkey station was the research work of the outstanding Russian biologist Ilya Ivanov, the originator of the modern method of artificial insemination in animal breeding. Before the Revolution Ivanov submitted to Prince A. Oldenburgsky, trustee of the Institute for Experimental Medicine, a plan for cross-breeding anthropoids, as well as anthropoids and man, by artificial insemination. Ivanov believed such work could help obtain valuable new ape hybrids, and also serve as one of the ways of demonstrating the degree of phylogenetic proximity of man and the higher apes. Later in a report to the Commissar of Public Education A. Lunacharsky Ivanov wrote that fear of the Holy Synod proved to be stronger than the desire to support this initiative. The tsarist censorship would not allow this scientific project to be even mentioned in the press.

However, after the revolution was successful and Soviet rule established, Ivanov was told to go ahead. "To carry out cross-breeding of the higher primates," says the book, "the biologist was sent abroad several times, Africa being one of the countries he visited."

In November 1926, Ivanov and his son arrived in Conakry, French Guinea, and set up a collecting station for primates to be sent to Sukhumi where the breeding station was being built. They soon learned that the methods used by the local Africans in catching chimpanzees left a good deal to be desired.

The Africans, armed with all sorts of weapons and accompanied by dogs, would surround a troop of chimpanzees, shout at them, set the dogs on them, and drive them up a tree. Then they built a big fire at the tree's base, and when the stifled apes jumped down to escape the fumes, the Africans stunned them with clubs. What with the dogs' attacks and the heavy fall, as well as the blows, most of the older animals were killed, and even the young ones were usually injured. Ivanov insisted on using a different method of capture, throwing nets over the apes. He got fewer of them after that, and they were mostly young ones, but at least not so many were murdered in the taking. The expedition could not be called a success, even so. Amoebic dysentery ran through the collection of monkeys and apes the Ivanovs had so laboriously gathered, and they lost ten young chimps to the epidemic before it could be halted. Travelling conditions were arduous, too: The older Ivanov had heart attacks, and the younger one got malaria.

"Thus, of the 30 anthropoids planned for the Sukhumi station, they could get together only 10, as well as two baboons. Apart from this number, three anthropoids were being used in Ivanov's experiment," say our authors, and in another moment we have the story:

> In early July 1927, the whole batch of monkeys, including three chimpanzee females inseminated with human sperm, were loaded in Conakry on a French ship bound for Marseilles, where they were to change ships. Ivanov intended to continue his experiments in Sukhumi. The heavy rolling of the ship, the heat and the gazing crowds round the cages affected the animals' condition. The situation was aggravated by an outbreak of dysentery with fatal cases among the 700 African soldiers on board. At that stage of the voyage the Ivanovs lost two anthropoids. In Marseilles, the Ivanovs were compelled by their illness to leave the monkeys which were temporarily accommodated in the Zoological Garden. . . .
>
> On August 24, 1927, 4 monkeys, including 2 baboons were delivered in Sukhumi, the rest, including Ivanov's three test chimpanzees having perished on the way. . . . On September 11 and 17, 1927, the remaining African chimpanzees died."

Ada Watterson Yerkes in her *Yale in Florida; 1929-1939,* gives an illuminating interim report of how things were at the latter date at Orange Park. The animals were thriving, she said, and "lead their individual and communal lives subject to the drives and limitations, the laws and the possibilities which heretofore have been known, if at all, only to themselves, but of which investigators are beginning to get teasing glimpses. Of the first four animals who roamed our New Hampshire pastures in their childhood two remain; Dwina died in childbirth and Bill of pneumonia. Pan, rusty-backed, lusty male, obligingly stages for visitors a dramatic dance of pounding, banging, hand-clapping rhythm around and around his big open-air enclosure. . . . Amongst the other animals the huge matriarch Mona and her tiny daughter Mu, born in April 1939, are at present oldest and youngest of the colony. Mona is approaching thirty years of age. She is the mother of Cuba who accompanied her from Havana, and of the twins Tom and Helene. . . . She is the only parent who voluntarily holds her newborn baby out for observation and allows human friends to touch its hands and feet."

Her husband does some summing up of his own in *Creating a Chimpanzee Colony:* "Among the personal rewards of association with chimpanzees as experimental animals there comes to my mind first of all the discovery that they have an appreciation of kindness and fair play . . . which may be compared with the trust of a human child. . . . Downright honesty is the only safe basis for mutual confidence between ape and man, as between men. This discovery is highly significant because what I have seen represents, if not the beginnings of human conscience and morality, the intimation or dawn of a comparable development in another type of primate. . . . We did much during the first decade of our work to extend and increase knowledge of chimpanzee life and to contribute to the solution of varied biological problems, but the sum seems trivial in comparison with what should be known—a mere beginning."

Though I admit that the account of Lapin and Fridman about the early days of Sukhumi does not sound like a success story, it did turn into one ultimately. Today, after many false starts and failures, the Sukhumi colony of hamadryas baboons is flourishing, as are other species of monkey. Anthropoid apes, however, are conspicuously absent, and crossbreeding experiments involving humans seem to have been abandoned. But, as we know, the story dies hard.

3

Primates in Demand

Long before Robert Mearns Yerkes died in 1956, his colleagues throughout the world were envious of the chimpanzee-breeding colony that he founded at Orange Park. By 1950, biomedical and psychological research workers in America were nervous, for the supply of subhuman primates most used in laboratories, the *Macaca mulatta,* or rhesus macaque, of India, was threatened. That the researchers themselves had contributed to the incipient shortage was no comfort, but it was the fact. Having discovered that for many of their purposes primate subjects are more satisfactory than the commoner (and cheaper) laboratory animals, such as dogs, cats, and rats, the scientists felt that there was no turning back, and one sees their point. In reproductive physiology, for example, no animal bears as close a resemblance to the human being as the anthropoid ape or, to various lesser degrees, the cheaper and more plentiful monkey; and just now, with everyone preoccupied with the question of population, reproductive physiology is even more important a science than it was in the past. There are also the nervous and cardiovascular systems, certain infectious diseases common to man and monkey, and many aspects of behavior among the infrahuman primates that shed light on our psychological reactions and behavior. In endocrinology, nutritional studies, dental studies, and others, rats and dogs prove to be comparatively distantly related to us, but the primate is our cousin.

Around 1950, the primate most in demand for laboratory use was the rhesus, not so much because it is the best model as that science had more or less started with it, and its anatomy and temperament were well known. A hardy, light-brown, long-tailed animal—on the average the rhesus weighs twenty-five pounds—it adapts itself to many terrains and diets. In *Mammals of the World,* the zoologist Ernest P. Walker first describes the whole genus of macaques—there are many species of them:

> Macaques usually inhabit forests, but some frequent treeless cliffs and rocky areas, and others are found in mangrove swamps. . . . The members of this genus are agile both in trees and on the ground and swim well. They are mainly diurnal, and feed on a wide variety of plant and animal food. . . .

He next discusses the rhesus macaque in particular:

> The rhesus monkey of India, held sacred by the Hindus, is the best known of this group as it is extensively used in biological laboratories for experimental purposes. . . . These are the first monkeys to be shot into the stratosphere in rockets. The Rh blood factor was first demonstrated in rhesus monkeys.

Hence, in fact, the symbol itself—Rh factor.

Because Hindus do not kill or injure rhesus monkeys—or any other kind of monkey, for that matter—the animals used to flourish unchecked, not only in the wild but also near villages and cities, swarming over tilled fields to steal crops, entering houses and temples and shops to help themselves to whatever they liked, and even—as I once witnessed in Uttar Pradesh—meeting trains regularly at the station and extorting alms from passengers sitting in the carriages, mother monkeys holding their infants up to the windows as if in supplication but actually so that the little ones could reach the bars and get inside, where pickings were better. Other troops of rhesus attached themselves to special villages and depended on the inhabitants for daily handouts. Yet others continued to prefer a really wild life to such a suburban existence.

The Indian peasant, while holding the rhesus sacred, as Ernest Walker says, does not necessarily love him. The rhesus is, after all, an added burden on the poor—sacred but a burden. I recall the puzzled, frustrated expression on the face of a brisk British official at a meeting of Hindu agriculturalists when, having asked cheerfully, "Just what method do you propose to rid yourself of these pests?" he received for reply only embarrassed titters. The Hindus would not, could not kill their monkeys, but they did perceive a way out when *Macaca mulatta* became so mysteriously popular abroad. They saw that they could make money and achieve relief at the same time, without sin, by trapping monkeys and selling them unhurt to foreign buyers. Once out of their hands, it was no concern of the Hindus what might happen to them abroad:

> Thou shalt not kill, but needst not strive
> Officiously to keep alive.

At first this enthusiastic export of rhesus did not make any visible dent on the monkey population. In 1953, for instance, sixteen thousand of them—no great number—were caught and shipped away. But then in 1954, when work on the Salk vaccine was at its height in America, the number of exported animals jumped to sixty-four thousand, and sporadic objections that had long been heard now swelled to an outcry. Not only antivivisectionists both in America and Britain, but the comparatively moderate British Royal Society for the Prevention of Cruelty to Animals joined in.

The RSPCA knew what it was talking about, for its officials, who keep a close watch on railways, docks, and airports to make sure that animals in transit do not suffer, had disturbing complaints to make about the monkey traffic. They reported that many rhesus being transported by air and sea from India to America turned up dead on arrival in Britain where they were to be transshipped. The number of such deaths—from overcrowding, malnutrition, or perhaps simply terror—sometimes added up to 30 percent, and on one notorious occasion, out of a badly overloaded crate of young rhesus landed by air at Heathrow, not one had survived.

Nevertheless, apart from such humanitarian protests, there was little reaction from public or dealers. Rhesus still poured into America by the thousand—until the price began to rise and a monkey of a size and condition that would once have commanded the sum of twenty-five dollars now cost seventy-five or even more. This increase hurt laboratory budgets, and the workers in the labs were dismayed and indignant. How could there be a shortage, as the middlemen claimed? There had always been plenty of rhesus in India.

In fact the stepped-up demand was at last making a difference in the rhesus population, even before a greater threat appeared. Religious purists among the Hindus of India found themselves more worried about the subject than the farmers were, and they now joined forces with the RSPCA and other societies abroad that protested against the use of *Macaca mulatta* in research. The *Statesman,* an English-language periodical in India, reported on February 5, 1955, from London, that a deputation from British animal protection societies had asked the Indian government as represented by Mrs. Pandit, the High Commissioner for India in London, to prohibit the export of monkeys:

> The deputation, led by Mr. Peter Freeman, a labor member of Parliament and a member of the Council of the Royal Society for the Prevention of Cruelty to Animals, was cordially received.
>
> They pointed out that in the last 2 years 100,000 monkeys have passed through London airport for the United States of America. Of them 75 percent were intended for experimental purposes, the deputation alleged, and the remainder for use in rocket research.
>
> After the meeting an RSPCA spokesman said that although they had provided a hostel at the airport to give the creatures some comfort on their journey, casualties in the 2 years had amounted to about 1,000.

In the same issue was a related news story:

> ONE HUNDRED AND FIFTY MONKEYS SEIZED IN DELHI—
> CASE FILED AGAINST FIRM.
>
> About 150 monkeys were seized by the SPCA, Delhi, from a place on Bela Road on Friday. The monkeys belonged to the proprietor of a New Delhi firm and were packed in five crates, which, according to the SPCA, were below standard. . . . The SPCA stated that the crates did not conform to the Indian Tariff Rules which prescribed dimensions and sizes of crates for transporting monkeys. . . . The monkeys are earmarked for export and they were brought from Lucknow.

Similar cases were reported in the following weeks, and fines were imposed on one or two traders. Then on March 11 the *Statesman* carried a slightly different sort of monkey story. Headed BAN ON EXPORT OF MONKEYS—GOVERNMENT'S PRIOR APPROVAL NEEDED, it reported that the government of India had that day announced a total ban on the export of monkeys, except with its prior approval. "The sole cause of this step," the paper said, "is the recent accident in London in which some monkeys were suffocated to death. The exceptional nature of the accident has not deterred the Government though the monkeys were employed purely for medical research, especially on polio, from which hundreds suffer in India."

The ban was announced in the Rajya Sabha (Parliament) by Mr. T. T. Krishnamachari, Minister for Commerce and Industry, and in the debate that ensued some MPs, who represented people in the export trade, argued against the decision. Dr. Homi J. Bhabha, brilliant head of the Indian atomic energy commission, was cited as saying that monkeys were not being used in atomic research in America: the recent increase in export numbers, he explained, owed itself to the fact that the National Foundation for Infantile Paralysis in the United States had set about finding a vaccine for polio.

Dr. Bhabha's words had some effect on the Minister for Commerce and Industry, and so did the expostulation of many of the world's scientists. As a result, not very long after the ban was imposed, in July 1955, he relented enough to make a preliminary agreement on behalf of India with the United States, to permit the export of rhesus monkeys for medical research and the production of antipoliomyelitic vaccine. One of the conditions of this agreement was that the United States Public Health Service certify that these purposes were the *only* ones for which the monkeys would be used. It was also stipulated that the American requirements be evaluated by a National Advisory Committee, and that humane care and use of the monkeys be guaranteed. Both stipulations were quickly put into effect, and a year later, in July 1956, the Public Health Service received assurance that India would not interrupt the export of rhesus as long as the service continued its certification-inspection procedure.

All over the United States, research scientists heaved a sigh of relief. It had been a near thing. But even now all was not well. The supply of rhesus had fallen off since the ban, and the scientists were still worried. Even before the crisis, many of them had become concerned over the threatened shortage, and had begun to look around for some other species of primate that might take the place of *Macaca mulatta*. At first they considered only other species of macaque—possibly the Japanese, the bonnet, the stump-tailed or the pig-tailed. But there was no reason, apart from the fact that rhesus were so familiar, for insisting on that particular genus.

Some workers were already involved in studies of baboons, which offered interesting possibilities in serology and behavior if nothing else. And the monkeys of South America, including the cebus (the organ-grinder's favorite capuchin), and of Central America, though not as closely related as are Old World primates to Homo sapiens, would undoubtedly be useful in many disciplines, nevertheless. The fact remained, however, that none of these species was as numerous in the wild as *Macaca mulatta* had been in the good old days, and wouldn't stand a chance against the unlimited hunting that had almost done for the rhesus. The facts were stark and uncompromising: the scientific world must be more careful with its resources and curb its appetite for primates, or it would certainly wipe out one species after another. Logic dictated conservation in the wild combined with rationing, increased care in maintenance, and the encouragement of breeding in captivity. Unless research was to stop short, these methods offered the only answer.

The picture was clear, but to put a comprehensive program, however necessary, into action was not easy. Already for some years primatologists had been trying to interest government authorities in the problems they faced in attempting to get hold of experimental subjects, but the cause seemed remote to most officials and held little interest for them. It was obvious that they must be educated, but biomedical research men are not necessarily adept at lobbying. One of the scientists involved was Dr. Willard H. Eyestone, head of the Animal Resources Branch, Division of Research Resources of the National Institutes of Health in Bethesda—which branch owes its existence, as a matter of fact, to this very campaign.

"Serious consideration for more extensive use of nonhuman primates in biomedical research by American scientists gained strong impetus in the late 1940's," writes Dr. Eyestone. "Though several productive programs had been under way for about twenty years, the lower vertebrate species were used predominantly. The expense of building suitable facilities and of purchasing primates doubtless limited this use of these animals. The advent of specific sums of money from the Federal Government for primate research was yet to come."

The scientists made their first tries to get government money for primate research in 1947 and 1949, through the Division of Research Grants—DRG—which recommended the establishment and financing of a program for the procurement of chimpanzees for medical research, "to make available to all researchers in this country an adequate supply of chimpanzees." (I am now quoting from a publication of the Department of Health, Education, and Welfare—HEW—entitled *Regional Primate Research Centers: The Creation of a Program.*) There was no favorable reaction to DRG's two attempts at that time; the brochure's author puts the failure down to DRG's low position in Washington's pecking order.

The next try, in 1953–54, was made by a different body, a subcommittee of the NIH Committee on Radiation Studies, which proposed the creation of a primate colony for use in a national long-term primate radiation program. This proposal likewise failed.

In January 1955—no doubt nervous of the approaching parliamentary action in New Delhi—the Committee on Radiation Studies tried again, proposing support for several already existing primate laboratories in the United States. A few months later a Public Health Service memorandum reported that a survey had been made by that body of "the actual need for rhesus monkeys for medical and biological purposes." Now various other interested bodies moved in on the act, of which the most important was the newly formed National Advisory Committee on Rhesus Monkey

Requirements—NACRMR, of course—which repeated the recommendation for support of existing monkey colonies "to facilitate research": in April 1957, the NACRMR again recommended that the NIH think the matter over. All this activity was in vain.

"It would seem," the brochure's author coldly observes, "that concern on the part of a study section and a council was not enough. To create such a program it was necessary to have concerted action on the part of a national advisory council and institute staff on the one hand, and study section and DRG staff on the other. Without strong concerted efforts by a national advisory council and institute staff (or the equivalent) the delivery of a new program conceived at a study section level was extremely difficult or impossible. This is particularly the case when no interest in or enthusiasm for the program exists at higher administration levels." In other words Dr. James A. Shannon, director of NIH at that time, wasn't keen on primate colonies. One might say that the government in general was not overanxious, at that time, to subsidize government research.

But change was on the way. Early in 1956 two American doctors went —separately—to Russia, and both of them, at different times, visited the Russian primate breeding station at Sukhumi on the Black Sea. One of the Americans, Dr. James Watt, was director of the National Heart Institute of the NIH; the other was Dr. F. K. Meyer, director of the George William Hooper Foundation Research Laboratory of the University of California, San Francisco. Learning that their Russian colleagues at Sukhumi were currently working on hypertension in subhuman primates, both visitors were filled with the desire to emulate this kind of work at home. Dr. Meyer wrote enthusiastically about the Russian setup to the Heart Council, while Dr. Watt—whose travelling companion, as it happened, was Dr. Paul Dudley White—came home full of ideas for a great American primate research center something like Sukhumi, but more like a kind of primatologist's Woods Hole, the famous central oceanographic laboratory on Cape Cod. Dr. Watt and Dr. Meyer between them provided considerable impetus. Early in 1957, Dr. Watt discussed with the Heart Council and a planning committee a recommendation that the primate colony he envisaged be developed near some unspecified university, which would serve as the site of "a long-term, multiuniversity approach to cardiovascular problems." Naturally, as a specialist, he thought only in terms of heart research.

Now, at last, things started to roll. The Heart Council was prestigious enough to impress even Dr. Shannon, and soon the usual committees and subcommittees proliferated. These set up plans for the primate center, modified them, threw them out, and set up new ones. As months elapsed, the idea became more precise and detailed, but even so, as late as March 1959, the designed colony was still referred to as "the Cardiovascular Primate Research Station." At that point the Planning Committee got an important message from Dr. Shannon, who announced that he preferred to think of America's Sukhumi not as one big central Woods Hole for heart research, but as a number of smaller centers distributed throughout the United States—"regional centers." Dr. Shannon was interested at the time in another project for regional centers—of computers and nonprimate animal resources: his mind was working regionally, as it were, and the NIH committeemen saw the point. Such a suggestion was likely to appeal to the Senate Appropriations Committee more than the original proposition of one central station would have done. From the taxpayer's point of view a distribution of smaller primate colonies was preferable, because each would presumably bring with it prestige and job opportunities as well as grants. At any rate, in the fiscal year 1960 (the autumn of 1959) Congress voted the first appropriation, of two million dollars, for the project.

Though the Appropriations Committee approved the multiple-center idea, even that late in the day its members seem to have exclusively considered the matter in the light of cardiovascular research, for the centers were spoken of as "colonies which would permit heart disease to be studied over the lifespan of the animals." Fortunately for the other branches of biomedical science, the actual wording of the appropriation bill was couched in more generous terms, permitting that the funds be used for general research on primates. It was agreed that each center be located near a university called the "host"; and the financing was to be provided by Public Health Service grants awarded to the host university administrations. Each award was made up of two grants: one to meet construction costs including the land on which the center was built; and the other for total or base operation costs over the next seven years in most cases, though a few of the original base grants were actually for ten years. These stipulated time limits were the cause of bitter complaint from scientists hoping to work at the centers. How could a worker do anything good on life-span research or any other long-term problem, they asked, within so short a space of time? But it is a stubborn fact of democracy that the expenditure of public money must be checked at intervals, especially just before elections, and on this point the senators—no doubt, reflecting that their own professional life-spans might well be shorter than seven years—were obdurate. Most programs, they might well have pointed out, are funded on a year-to-year basis.

The planners outlined in detail the Siamese-twin relationship that they felt should be the pattern between primate center and host university. In the usual government language they said that each center was to perform both a local and a regional, or possibly a national, function. "Local function," they explained, meant that research programs should be carried out by the professional staff permanently connected with the center, working in cooperation with people from the various departments of the host university or medical center, and financed either directly by the government or by some special mechanism in which the university was involved. Each center was to have a director, and the research staff under him would have, "ideally at least," university appointments. The regional or national function was to provide—on request—space and certain basic equipment to visiting scientists from the region, the nation, and even outside the nation. However, during the recent economy drive of the Nixon Administration the last provision, which offered hospitality to foreign research workers, was cancelled: scientists from abroad must now get special permission to work in American primate centers.

Where were the centers to be located? The burden of making this decision was shouldered by the NACRMR, now—much to everyone's relief—rechristened the Primate Research Study Section. PRSS members travelled throughout the United States to inspect all the sites that were deemed eligible, at the same time spreading word that applications from would-be host institutions were in order. These applications promptly poured in, and dispute filled the air, for it seemed that everyone wanted a primate center. Nerves were frayed and friendships wrecked by the time the final decisions were made. As the annalist says, the main problem was that each decision "had to be based on a number of considerations other than the scientific merit of the primate research already being carried on . . . at the site selected"—in other words, political expediency. The members of the Study Section and the Heart Council wanted to ignore all political aspects of the deal and make their decisions on scientific merit and nothing else, but the United States is a democracy, and there it was. The Heart Institute staff, being in closer touch with the harsh facts of life and the Senate Appropriations Committee, were under considerable pressure to take the other point of view. At last, after "earnest and persistent persuasion" by the Heart Institute director, the Study Section was persuaded to accept a compromise solution. Geographically speaking the distribution of the Regional Primate Centers is hardly ideal, being definitely lopsided, but as there were only seven centers to be scattered through fifty states, somebody had to go short.

This is how it worked out:

Center	Date of Initial Grant	Dedication Date
Beaverton, Oregon	May 1, 1960	May 1962
Seattle, Washington	June 1, 1961	December 1964
Madison, Wisconsin	June 1, 1961	April 1964
Atlanta, Georgia	June 26, 1961	October 1965
Covington, Louisiana	June 1, 1962	November 1964
Southborough, Massachusetts	June 1, 1962	October 1966
Davis, California	June 1, 1962	September 1965

Thus we have the centers of Oregon and Washington next door to each other, as are Georgia and Louisiana, but Wisconsin, Massachusetts, and California are relatively isolated. Strictly speaking, the center at Davis is not regional, but national. Its host institution used to be known chiefly as a "cow college" or veterinary school. Today its curriculum is diversified, but as the one and only *National* Primate Research Center it still reflects the original interests of Davis and concentrates on problems of husbandry in dealing with primates.

In addition to giving support to the seven main centers, the Division of Research Resources keeps a friendly eye on other primate study centers or colonies within the United States. Occasionally it provides grants for these institutions, and in return the centers sometimes borrow animals from them. Among others these include the Southwest Foundation of San Antonio, chiefly known for its baboonery; the research laboratory at Holloman Air Force Base at Alamogordo, New Mexico, which has a large number of chimpanzees and some few of other species; the monkey island colonies of Puerto Rico; and the primate laboratory of the New York University Hospital near Tuxedo, New York. There are other concentrations of primates here and there in America, and conscientious primatologists can keep up with developments in their special world on most of these fronts through the *Laboratory Primate Newsletter,* published by the psychology department of Brown University and supported by a grant from the Animal Resources Branch, Division of Research Resources. There is also the *Primate News,* published by the Oregon Center, and *Current Primate References,* a publication of the Primate Information Center of the Primate Research Center at the University of Washington—not to mention *Mainly Monkeys,* published by the Wisconsin Center.

Certain choices of places for centers, such as that of the Davis campus for monkey breeding, seem logical even to the uninitiated. Obviously the

University of Wisconsin at Madison was selected because Dr. Harry F. Harlow, who immediately became the center's director, was already a member of the university faculty, and his work on the maternal affectional system of rhesus monkeys, especially the reactions of infant rhesus deprived of mother love, published by the Wisconsin Department of Psychology Primate Laboratory, was famous even among laymen. Many such recall an article about this research in *Scientific American,* illustrated with photographs of baby monkeys clinging to surrogate mothers made of terry towelling, which had what is called universal appeal.

When I started to visit primate centers, I was uncertain where to begin. Wisconsin and the infant rhesus sounded tempting, but in the end I decided that the logical place to use as diving board was the Yerkes Primate Research Center at Emory University in Georgia, just outside Atlanta, for it is, in a way, the direct heir of Dr. Yerkes himself. Yale had continued to serve as trustee of the Orange Park laboratory after Yerkes retired from the directorship in 1941, but as time went on and the original members of the Orange Park staff died, the Yale authorities found the responsibility for housing all those chimpanzees, so far from New Haven, burdensome. For this reason they had welcomed the offer made by Emory University to take over the chimp colony, and in 1956, the year of Yerkes' death, the transfer was made, Yale presenting the entire collection to Emory as a gift.

Georgia's climate is warm and mild, like Florida's, but the colony's new owners couldn't move the apes across the state line because they had no facilities for housing anthropoids and no money to build any. The primates therefore remained in Orange Park for a number of years, tended by Emory employees and visited by Emory researchers—both staff and students—who travelled over from Georgia to make their observations. This arrangement was very unsatisfactory, and it came as a great relief when the Primate Research Study Section accepted Emory's application to be named host for one of the new primate centers.

On the plane from New York to Atlanta, I read what Dr. Eyestone had to say about Emory and the Yerkes Regional Primate Research Center, as it has been named: "In keeping with the tradition of the Yerkes laboratories, the center is largely devoted to studies on anthropoids. The behavioral sciences dominate the present research program. Other areas include anatomical and physiological studies of the brain and on blood fractions relating to immunological responses. Normal studies of greater apes at this center will be used to establish their biological 'profile,' since so little is known about them."

Spring was well advanced outside Atlanta, especially at Emory where my taxi followed a winding road through a mile of woods, with here and there a flowering shrub. The taxi driver talked without stopping in an almost completely incomprehensible language, until we suddenly arrived at a clearing, and he pulled up at the foot of a shallow flight of stairs that led up to wide glass doors set in an imposing, new-looking building. Like four other of the centers, Yerkes seems a long way from its host campus, and I wondered if this was an example of government tact. At any rate, as I paid the driver, he looked suspiciously at the center's sign in metal letters on the stone balustrade, and shook his head doubtfully. I heard later that there had been a certain amount of protest from the townspeople when they got the news that large numbers of great apes were coming to live nearby, but from where I stood on the driveway among parked cars and a small university bus, there was no sign of any but human primates.

Indoors, in a pleasant anteroom, the receptionist's window bore a sign: "Visitors may only be shown through the Center by prior arrangement." After using the telephone to check up on me, she asked me to wait. What appeared to be a university party was there ahead of me, comprising one girl carrying a camera and a number of youths with notebooks. The girl sighed noisily, looked at her watch, looked at the one door that led into the rest of the building, and asked a boy, "Is that locked, do you think?"

It was the receptionist who answered, in kind, firm tones: "*All* our doors are kept locked."

A jingly sound was heard; the inner door opened; and a man in a white coat came out and gathered up the party. The door promptly swung shut behind him, so that he had to unlock it all over again to lead his charges through. Alone in the waiting-room, I sauntered around and looked at a bronze bust on a pedestal in a corner, labelled

DR. ROBERT M. YERKES, 1876–1956
First Director, 1929–1941

Along the wall was a glass case holding a skeleton of what the plaque told me was Mamo, a female chimpanzee captured in French Guinea in 1930, one of those sixteen apes sent by the Pasteur Institute to the Orange Park station when it opened. Mamo seems to have died on April 5, 1941, at the estimated age of thirteen and a weight of 42.8 kilograms (94⅓ pounds). Not a great age for a chimp, I reflected, and wondered what had killed her. Then the door clicked and jingled, and someone came to lead me, too, past the inner wall.

I was to visit Yerkes until its wide halls and laboratories became familiar, and meet many people who worked there, but my introduction that day was to Major General George T. Duncan, retired, an assistant director, as he said, on the administration side, and—though he didn't say so —obviously a good man at public relations.

"It was getting harder and harder to keep our end up over in Florida before 1960," he said. "We took control in 1956, and costs began climbing right away. Things were all the worse because under the arrangement they had then the chimps were the state senate's responsibility, and senate support was—well, let's say inadequate. Sometimes we had trouble even meeting the food bills. The chimps knew hard times. Then at last we got the grants, and what a relief *that* was. After that, we were all pretty busy with architects and engineers and contractors, building new quarters here in Emory for the great apes. Most of the other centers haven't the same problems as ours because their animals aren't so big; we need lots of room and lots of strength in the building. You've only seen the building from the front so far—I think you'll be surprised when you find out how far we extend to the rear.

"Anyway, finally everything was ready, and high time, too. We started moving in 1963, but it was '65 before all the primates got here. It was awfully complicated. Some of those animals hadn't been shifted in twenty years at least; some had been there for thirty; and it was naturally upsetting for them. We used specially built trucks, and they were driven slowly and carefully, I can tell you. There were travelling cages and a flotilla of cars. Ours is the most valuable collection of great apes in the world, and it's becoming more valuable the whole time because of what's happening in the wild."

The major general stared out the window as if he could see the wild threatening the center's very threshold. "Take orangs," he said. "In their natural surroundings they're dying out. But we have a good breeding colony of orangs right here—thirty-five of them—besides fifteen gorillas and eighty-five chimpanzees. As for small primates, we've got—let's see now, I've got it all written down somewhere—oh, yes, here we are: seven hundred animals in the smaller primate collection, which represents twenty-five species."

I asked if this wasn't a fairly new departure for Yerkes, which had always concentrated on chimpanzees, and he nodded.

"Oh, yes, we've diversified since we took over from Orange Park. The staff here wants smaller animals: we can't afford to use anthropoids all the time for a lot of the work we're doing nowadays. In Yerkes' time—I mean the old man—and for a while after he retired as well, practically all our studies were behavioral, with a few exceptions—reproductive habits, for instance, and general physiology now and then. But when we got the word that this was going to be a regional center, and we had a grant for seven years ahead, we expanded: we started getting monkeys. For instance Geoffrey Bourne, our director, is working with Dr. Nelly Golarz on the presence of pathological changes in muscles of primates and also the effect of weightlessness in space on muscles, using monkeys. We used to have only five persons on the faculty, with a staff of 35—that was in Florida—but now we have 23 faculty and about 120 staff. And there's a subordinate station out at Lawrenceville not far away, where some chimpanzees and other animals are to be kept in what approximates natural surroundings—though, of course, they'll be confined in some way. The idea is to watch their behavior there in conditions of comparative freedom, to see if it's any different."

He pulled out a chart like a floor plan, full of little boxes containing names and information, and began ticking off details on it. "Here we are. This at the side represents our veterinarian Norman B. Guilloud, in charge of Vet Medicine and Animal Welfare. He also oversees Surgery and Pharmacy, X Ray, Hematology, and Veterinary Pathology—that last is with Harold McClure. One of Guilloud's projects is the establishment of norms of baseline physiological data for the various species, such as body temperature and pulse rate. This has never been done comprehensively, and it's pretty important to know the normal temperature for a chimpanzee, for instance, or a pig-tailed monkey. McClure in Pathology is doing research on animal leukemia—very interesting; he works with our new electron microscope, which you'll have to see. He postmortems every animal on the place that dies, no matter what the ostensible cause of death. Here's Neurochemistry where they're working on enzymes and the neuroanatomy of primate brains. Charles Graham in Histology and Histopathology is a zoologist: he's trying to induce cervical cancer in the squirrel monkey.

"Now let's see what else is going on. . . . Maurice Sandler in Histochemistry and Dr. Nelly Golarz are on muscle diseases, and Dr. Golarz is working with the director on a project with NASA on the effects of long-term space travel. Gross Anatomy and Physical Anthropology—that's William Charles Osman Hill from England; he's writing a detailed study of primate anatomy and taxonomy and accumulating a bank of anatomical and embryological material. In Psychobiology, Richard Davenport and Charles Rogers are studying the behavior of animals in normal laboratory conditions, then under the influence of drugs—all the things we use ourselves, such as stimulants, depressants, tranquillizers, and alcohol.

"Now let's see what else—oh, yes, Immunology. Dr. Richard Metzgar and Dr. Sigler are working on blood typing and the isoantigens of the anthropoid. All that's becoming important now, since work on transplant tissues is speeding up. Who knows? One of these days they may be using animal organs from primates, kidneys for instance, in humans, and the rejection factor's got to be watched. Of course, primate blood's been looming large on the scene for a long time now, what with the Rh factor and all that. In the Neurophysiology lab Bryan Robinson is studying brain stimulation—telestimulation and telemetry. Then there's Neuroanatomy and Optic Physiology——" A man in a laboratory coat came in. "Oh, there you are, Guilloud," said the general, and cheerfully handed me over to a new guide.

I might have been swamped in this flood of information if it hadn't been for Dr. Eyestone, whose little book about the centers had prepared me for much of it. "Each of the seven centers has its own general area of investigation that is related to, but does not duplicate, research being conducted in the other centers," he had written. ". . . The central theme of each center's program developed largely around the interests of a particular institution submitting an application"; and a later commentator, one of those anonymous people who draw up reports for the authorities, has said, "Specifically, each Center has developed a 'research personality' whereby investigations are concentrated in specific areas of study." Apparently it was for me to distinguish between the Yerkes personality and those of the others. I braced myself.

Under the wing of Dr. Guilloud, a soft-spoken man, I visited the Yerkes laboratories, met doctors, and admired equipment, especially the new electron microscope. I peered in at research library and operating theater. Then my guide was called to the telephone, and as I waited for him in the corridor, I studied a series of placards that lined the walls. The centers, as I was to discover, are anxiously, almost obsessively educational, and these posters were clearly designed for somebody as ignorant as myself. A series of photographs with text, dealing with a case of epilepsy in a gorilla, was actually possible to understand, and so was the account of an air sac infection in an orang-utan, and the measures taken to cure it. Air sacs are those facial pouches that develop in the mature orang, giving his countenance a square, enormous look. The last line of this account read, triumphantly, "Uneventful recovery." I had started on a new series dealing with the sexual cycle of the chimpanzee, and had just reached the line, "Gestation period approximately 227 days in length," when Dr. Guilloud reappeared to say that he had coffee waiting in the office. All the primatologists I met during the months that followed seemed to drink gallons of coffee.

While we sipped from plastic cups, Dr. Guilloud told me how they had discovered that gorillas and orangs are vulnerable to polio. Nobody knew this until the disease attacked a gorilla at the Yerkes laboratories at Orange Park in 1964, just before the last of the animals was moved to

Georgia, when the center people were still going over to Florida to make their observations.

"He was a young animal named Bandam, a recent arrival," said Dr. Guilloud. "We could see that he didn't feel well, so we took his temperature, and it was very high. What with one thing and another, I thought he might have polio, but there wasn't any case on record of anthropoids getting the thing, and I couldn't be sure. We isolated him just the same, and I contacted an experienced pediatrician in Jacksonville who knows a lot about polio, and he was willing to have a look at samples, so I sent them off and waited—those tests take quite a while.

"Finally he called me up and diagnosed it; it was the real thing, all right. In any case after ten days there couldn't be any doubt about it because Bandam showed a perfectly typical paralysis. He survived. We've been treating him for it just the way you'd treat a child with polio—he gets massage every day, and therapy, and he moves around in a little walker: he manages it very well.

"But it wasn't all so happy an ending. It was in July he came down with it, and as I said, we isolated him right away, but he'd already handed it on. In September one of our orangs came down with polio, and so did another young gorilla which died. Now we immunize all the great apes with regular vaccine. Fortunately monkeys seem to be naturally immune—we haven't had to worry about them."

The tour resumed. I was permitted to don a laboratory coat and sit on a high stool in the apes' playroom in the nursery, where I watched a number of little chimpanzees first as they romped around, and then a contingent of small gorillas. They wrestled, raced, pulled toys back and forth, and now and then in passing gave me a friendly slap. Next we paid a visit to a large cage to look at an expectant chimpanzee mother. She stared straight past us as if we did not exist: she had more important matters on her mind.

"We keep a close eye on her," said Dr. Guilloud. "She's pretty old to be giving birth, not that age seems to worry chimps in these matters. As far as we know they don't go through menopause. We're not sure if any of the infrahuman primates do, but we'd like to be sure." (Since then, however, Dr. Gertrude van Wagenen has published the results of a study made on rhesus females, which indicates that these monkeys do indeed pass through something similar to menopause.) Dr. Guilloud added, "You know we still have one of Dr. Yerkes' original chimps—Wendy."

I was astonished. Wendy? I had read of her, of course; she was acquired in the very early days, along with a young male named Pan. The names themselves date the animals. It was like finding Mrs. Noah alive and well on Ararat.

"Yes, she's about forty-six," said Dr. Guilloud. "Of course, we can't be absolutely sure, since Dr. Yerkes himself didn't know exactly how old she was when he got her, but she can't be far off forty-six. She probably represents some kind of longevity record."

Accompanied by the now familiar jingling of keys, we passed through several doors and arrived out in the open—if one can call it that—far to the rear of the building, at one end of a long covered walk that led past one high, roomy cage after another. All down the line, with a soft, scurrying rustle, apes hurried to the front bars of their compartments to see who was coming and to watch us walk past. They sat on crossbars or stood clutching the bars that made up their front walls, pressing their noses between, staring. They obviously knew Dr. Guilloud, but were curious about me, and I thought they looked snobbish, superior, even insolent. I suppose their silence put me off. Once in prewar England I was taken to a tea party in South Kensington where nobody had ever seen a real live American before, and they all sat and stared at me in the same way. I still remember it sometimes in dreams, and wake up sweating, but things proved to be a little better than that among Yerkes' great apes. Dr. Guilloud took my arm, about halfway along, and propelled me toward the outside of the walk, away from the cage fronts, saying, "You'd better stay as far off as you can. Some of the chimps spit, and they have a good aim. Those two ahead are the worst."

One of the two he indicated waited until we were nearly out of reach before he let fly. The other didn't even try, but farther along, as we passed a smaller chimpanzee, the animal spat at Dr. Guilloud, who stopped short and regarded him in surprise.

"Why, you've never done that before!" he said in hurt tones, and then, suddenly, he spat back.

Outside an orang-utan's cage he paused and said, "We're really proud of this girl." She was a huge reddish-orange animal whose eyes as she crouched comfortably on the floor were almost level with ours. A lively little orang climbed over and around her bulk as if she really were the mountain she resembled, while the mother's intelligent red-brown eyes remained fixed on Dr. Guilloud's.

"We had to do a Caesarian to save her life," he said. "There was some risk for the baby, and it would have been a crime to lose him, but everything turned out fine. How's the boy, Bessie?" Gently he scratched her chest through the widely spaced bars. "He looks fine," he continued. "Does you credit. . . . You see we let them rear their own children if they don't reject them, and this one's a good mother."

The little ape sprang suddenly to a side bar and clambered up to a point above our heads, from where he blinked down at us. Dr. Guilloud had stopped scratching Bessie and was moving away when, with what looked like slow motion but couldn't have been, she grasped his hand and held on to it. She didn't do anything else, merely sat with her fingers wrapped around his, eyes fixed on his face. She was as immobile, as impassive, as impressive as any Englishwoman at a South Kensington tea party.

Mildly Dr. Guilloud said, "Let go, Bessie, that's the girl."

Bessie did not let go. She didn't do anything else either. I, too, did nothing: I just stood there. I felt that if Dr. Guilloud wanted me to do anything such as call out the guard, he would say so.

"Come on, Bessie," he said in even tones. "I can't stay here all day, you know that. I have other things to do. Come on, girl, let go."

When there was no reaction to this, he lifted his other hand—slowly—and began working at one of her fingers, uncurling it from its grasp. He talked as he worked, like a loquacious oyster opener: "That's the girl. That's it. That's the girl. A joke's a joke, isn't it, Bessie?" She did not resist, but she didn't help either. Now he had that finger prised up, and it hovered where he left it, like a forgotten sausage in a spaceship.

Dr. Guilloud calmly set to work on the next finger, forcing it up while Bessie continued to stare into his eyes. Having uncurled that one, he started on the third, which was also the last she was using. It seemed to let go more easily, and then he was free. He stepped back slowly, out of reach. "Good girl, Bessie," said Dr. Guilloud.

I asked urgently, "Are you hurt?"

Fingers to mouth, he shook his head. His words when they came were muffled. "Perfectly all right," he said cheerfully.

Bessie watched us go without comment or movement. The baby jumped to her back again and peered at us over her hairy shoulder.

4

Testing, Testing

A year later, when I revisited Yerkes, I was sorry to find that Dr. Guilloud had moved to the state university at Athens. There were other changes on the staff, too, for things never stand still among laboratory men. Dr. Bryan Robinson, for instance, had gone back into private practice; his work at the center was now being done by Dr. Adrian Perachio, working on behavioral problems.

"Though I'm really a neurophysiologist," he said. "Robinson's program, which he published a paper about, concentrated on remote control of the brain, or rather, the stimulation of certain areas of the brain, and so does ours. He worked on agonistic, or combative, and sexual behavior in the rhesus monkey: you may have seen a report on related work done by José Delgado at Yale. Well, one of the things that's been found out is that stimulation in the right area affects whatever quality it is—I don't know that we have a word for it—that leads a monkey to dominate his group."

"Leadership?" I suggested.

"Possibly. This dominance can be adversely affected by stimulation: that is, an animal that was dominant may become subservient when stimulated, and it's interesting to note that even after the stimulation ceases the animal may remain in that condition. The opposite effect can also be obtained: an animal that is usually humble—'naturally' humble, you might say—may become dominant and remain so.

"We have also discovered that stimulation can cause sexual behavior, or rather step it up, but then, after a period of time, say a week, on cessation of the stimulation, the sex behavior is suppressed, as if the animal were exhausted.

"We also work at things from the biochemical angle; we work with biogenic amines. Certain substances are enzyme inhibitors that modify the metabolism of the amino acids. Treatment with these inhibitors has disruptive effects on sleep patterns and eating; it also seems to change sex behavior. One of these substances is being used in hospitals as a special drug against cancer, for the following reason: it is known that some carcinoid tumors secrete a particular amine, and they figured that if the secretion were to be stopped the cancerous growth should cease. It does work like that, but there are side effects—you may have read reports of them. Use of the drug alters the patient's behavior, and there have been stories in the newspapers of how the patient becomes sexually stimulated."

"I thought that was another drug," I interrupted. "Something used for Parkinson's—L-dopa?"

He nodded. "Yes, that one has been reported to have similar side effects. Well, it was observed in the laboratory that the drug I spoke of evidently had this effect on rats and cats, and we decided to try it out on primates, too: the areas we stimulate are rich in this enzyme, and we felt justified in using the inhibitor. After all, you want as many handles as you can get. Well, what we discovered is interesting. The drug does *not* bring on hypersexuality in terms of the end point of sexual behavior. It increases mounting behavior and prolongs the time of intercourse but actually inhibits ejaculation. The published reports on the sexual effects of the drug on animals were in conflict with my data. To my mind, the real question is what its effects may be on the biochemistry and neural bases of the subject. We would like to find out."

Another set of investigations, Dr. Perachio said, was into the neural bases of the sleep-waking pattern, for which they were using three New World species of monkey, all in the same taxonomic family—the squirrel monkey (or saimiri), the cebus, and the owl monkey.

"The owl monkey is the only nocturnal simian that exists," he said, "though there are various nocturnal prosimians of the bush-baby or galago type, primates lower on the family tree than monkeys." They were studying the rapid eye movements that occur at certain stages of sleep, which evidently have some connection with the visual and auditory systems; these systems themselves are connected. "Reflexes from the inner ear are connected with the eye muscular system," he said. "Impulses move from the labyrinth of the inner ear to the eye muscles, and we're trying to track down the relationship."

They would like to know, he said, to what extent this eye-ear activity is correlated with waking-sleeping states, and they were recording movements in order to correlate the dependence on, or independence of, these action systems with each other. "It's a very complex problem. It's funded by NASA, who want to know among other things what if any are the long-term effects of rotary movement and drastic alteration of sleep patterns such as the astronauts undergo. In short, what will happen to the organism generally. Naturally they're much concerned. After all, astronauts are in a chronically abnormal vestibular pattern." He looked gloomy.

"The men themselves, you know, are for the most part resistant to these studies," he admitted. "Oh, well—Do you remember that monkey Bonny who died after a space shot? She was launched June 29, 1969, and they planned to leave her upstairs for thirty days, but they had to bring her down after only eight and a half days, and she died seven hours after landing. Remember the outcry caused by that affair? But Bonny was the most instrumented organism that's ever been put up. They collected more information from her flight and death than they've had from any of the other space shots. The reasons for her death haven't been released, but they're going to carry tremendous weight in future experiments. For myself, I think we ought to use more primates and not so many men."

Another change that I found at Yerkes was that the staff had been augmented by the psychobiologist Duane Rumbaugh, formerly professor of psychology at San Diego State and now associate director at Yerkes. The director, Dr. Geoffrey Bourne, explained that Dr. Rumbaugh had been working at the San Diego zoos with nonhuman primates for some years, and that he was especially interested in anthropoids.

"He's brought some gibbons with him, as a matter of fact," said Dr. Bourne.

As the English say, the penny dropped at that moment: I am an enthusiastic zoo visitor, and I recollected meeting Dr. Rumbaugh when he was giving intelligence tests to gibbons at the California zoo. Now Dr. Bourne and I ran him to earth in a laboratory, standing next to a mysterious machine like a big black box, a miniaturized version of the booth they used to use on TV for quiz shows, in which the contestants were supposed to go through agonies of thought before they gave their answers. Along the top of Dr. Rumbaugh's booth ran a band of tiny windows, and from inside came little noises—not utterances, but mechanical noises. Every so often this muted thump was accompanied by a light shining out of a little window. A white-coated girl sat by the box and took notes.

"We have a gibbon in there," explained Dr. Rumbaugh, "running through tests on a learning set. I started this work in 1960 as a result of Harlow's work at Wisconsin, and I'm using his methods—with certain adaptations. It's hard to explain, but the gibbon in there is looking at a series of slides, pictures of objects in an irregular sequence of repetition. One certain picture is the right answer: every time she sees that one she is supposed to press a lever and get a reward—something to eat."

I listened for the lever and watched the windows. *Ping,* light; *ping,* light; *ping,* no light; *ping,* light; . . .

"She's got it pretty well now," he said with satisfaction.

In 1949 Dr. Harlow and his wife Margaret published an article in *Scientific American* entitled "Learning to Think," from which I quote certain passages:

> The speed and complexity of a human being's mental processes, and the intricacy of the nerve mechanisms that presumably underlie them, suggest that the brain is not simply a passive network of communications but develops some kind of organization that facilitates learning and thinking. Whether such organizing principles exist has been a matter of considerable dispute. At one extreme, some modern psychologists deny that they do and describe learning as a mere trial-and-error process—a blind fumbling about until a solution accidentally appears. At the other extreme, there are psychologists who hold that people learn through an innate insight that reveals relationships to them.

The Harlows wrote that they were trying at Wisconsin to investigate these positions, and if possible to reconcile them, in a series of studies, some made with young children but most with monkeys.

> One of the first experiments was a simple discrimination test. The monkeys were confronted with a small board on which lay two objects different in color, size and shape. If a monkey picked up the correct object, it was rewarded by finding raisins or peanuts underneath. The position of the objects was shifted on the board in an irregular manner from trial to trial, and the trials were continued until the monkey learned to choose the correct object. The unusual feature of the experiment was that the test was repeated many times, with several hundred different pairs of objects. In other words, instead of training a monkey to solve a single problem, as had been done in most previous psychological work of this kind, we trained the animal on many problems, all of the same general type, but with varying kinds of objects.

What happened as the monkeys became adept was extremely interesting. At first the animals learned by "the slow, laborious, fumble-and-find process," but as a monkey got accustomed to the sort of problem facing it, it showed progressively greater efficiency. Eventually it showed "perfect insight" when faced with that particular kind of situation, and solved the problem in one trial: that is, if it chose the correct object the first time, it rarely made an error afterward. If it guessed wrong the first time; it immediately shifted to the correct object, and subsequently responded almost perfectly.

> Thus the test appeared to demonstrate that trial-and-error and insight are but two different phases of one long continuous process. They are not different capacities, but merely represent the orderly development of a learning and thinking process.

The Harlows ran a series of these "discrimination problems," as they called them, on a group of nursery school children two to five years old, rewarding those who succeeded with bright beads rather than raisins or peanuts. They reacted as the monkeys did, making many errors at first and gradually learning to solve a problem in one trial. As a group, say the Harlows, they learned more rapidly than the monkeys, but they made the same types of errors, and the smartest monkeys learned faster than the dullest children.

"We have called this process of progressive learning the formation of a 'learning set,'" say the authors. "The subject learns an organized set of habits that enables him to meet effectively each new problem of this particular kind." They add that a single set provides only limited aid in enabling an animal to adapt to an ever-changing environment, but a host of different learning sets may supply the raw material for human thinking. Such sets were made by complicating the original test, reversing the reward value of the objects once, for instance, and then going back to the first method, or using three objects to choose between instead of two. In some cases, the monkey had to learn that it was not the shape of the object that was important, but its relation to the other two.

> One of the most striking findings from these tests was that once the monkeys have formed these learning sets, they retain them for long periods and can use them appropriately as the occasion demands. After a lapse of a year or more, a monkey regains top efficiency in a few minutes or hours of practice, on a problem that it may have taken many weeks to master originally.

Thinking over this evidence, the Harlows decided to re-examine the evidence offered to support the theory that animals possess some innate insight that has nothing to do with learning. This theory is based primarily on Wolfgang Köhler's chimpanzees and his interpretation of their behavior regarding the sticks they used as tools, knocking down bananas with them, raking bananas in, and so on. "That the chimpanzees frequently solved these problems suddenly, as if by a flash of insight, impressed Köhler as evidence of an ability to reason independently of learning," say the authors. "He even suggested that this ability might differentiate apes and men from other animals."

But Köhler's observations had been made long ago, in the twenties, when the study of chimp psychology was all but unknown. All his animals had been captured in the forest, and, as the Harlows pointed out, he had no record of their previous learning. Nowadays, when we can watch chimps born under our eyes, as it were, things look rather different. Several people have tested young chimpanzees in this respect. The younger they are, the slower they prove in learning to use tools. None of them seem to solve their tasks at first with "sudden insight."

I had never heard how gibbons rate in these tests, and I asked Dr. Rumbaugh, "How do they do—do they catch on quickly?"

He shook his head—sadly, I thought. "They're notoriously slow by the accepted methods of testing. Here, let me show you a graph that gives comparisons. This curve represents the work done by a chimpanzee, plotted with the coefficients of date and performance. You can see here, down in the corner, at the beginning of the series, that she was fumbling, and for several days the line stays on a level. Then suddenly—see?—it turns up; the curve goes on pretty regularly until she's making practically perfect scores all the time. In other words, she's caught on: she has that learning set. And this curve is for rhesus, which follows the chimpanzee performance almost exactly. Sometimes rhesus are even quicker than chimps. But here, look at this record." His finger traced a line that stayed close to the baseline, day after day. Even after it had lifted for a little, it slumped again. "And that, I am afraid, is for the gibbons," said Dr. Rumbaugh.

I was sorry about it. I am very fond of gibbons. I know they haven't the same sense of logic—if one can call it that—as chimpanzees or monkeys, but I have never for a moment thought them stupid. Clearly Dr. Rumbaugh agreed with me.

"I'm not satisfied that we've been testing them properly," he said. "We've had the same sort of difficulty with squirrel monkeys, and I've devised a new method that promises to give better results for both species. I reached the conclusion that their attention is distracted, during these tests, by irrelevant visual clues in the apparatus. Let's go upstairs to see the ordinary apparatus that we use with chimps."

On the top floor, where a large number of apes were housed, a chimpanzee was being invited to play something very like the old thimble game. A young girl sat at an elaborate kind of table on wheels in front of the chimp's cage. Erected on the table, at the cage side, was a framework through which, as through a window, she and the chimp regarded each other as long as the window was unshuttered. It could be shuttered, however, by a guillotine arrangement operated by the girl. With the window closed, she arranged two cups of different colors, placing each upside down with one covering a few raisins. Then up went the shutter, and the ape, reaching eagerly through the bars, took the cup off the raisins and grabbed them. The girl made a note on a pad, closed the window, and rearranged the game. During the whole time we watched, the animal made almost no errors.

"Well, that's the general idea," said Dr. Rumbaugh, "but it doesn't work with gibbons because they are so easily distracted by other things on the apparatus. Let's go back to the office."

There he embarked on a description of the modifications he has put into a learning set formula devised at Wisconsin by Harlow, by adding something called Transfer Index. Later I read an article of his in which he advanced an interesting possibility:

> Gibbons and squirrel monkeys are highly arboreal, only rarely coming to the ground for other than very short intervals. Of the large apes, the orang-utan is the most arboreal, whereas the chimpanzee is intermediate in this regard, being somewhat less arboreal than the orang-utan but less terrestrial than the gorilla. A working hypothesis might be that associated with arborealness in primates is selection for readiness to attend to immediate foreground clues, which in the trees define routes of locomotion. This characteristic, which might complement life in the trees, could in formal test situations profoundly distract attention by allowing relevant clues to become effectively embedded or otherwise obscured by irrelevant clues, particularly of those planes which are more proximal to the eye than those which provide the reliable and relevant clues of the task. That they probably do differ in this regard makes comparative studies of primate learning, where widely disparate genera and/or age levels are studied, a particularly problematical endeavor if data are evaluated and interpreted via the traditional method of direct comparison.

Back in the office, over coffee, glancing through a magazine connected with the center, I suddenly recognized a name.

"Look here," I said, with some excitement, "you never told me you'd brought *Gabrielle* from San Diego with the other gibbons. Why, I knew Gabrielle when she was so high—in the animal nursery at the zoo. She was an awfully nervous baby. How is she, anyway?"

Dr. Rumbaugh looked pleased. "You knew her, did you? Yes, she *was* a neurotic little thing, but she's grown up to be a beautiful animal. We have her in the country here. Well, well. If you knew her, I think you might be interested in something I've found out about her."

He proceeded to tell me. I ought to say, perhaps, that Gabrielle is a hybrid, the daughter of a lar and a Moloch gibbon. As an infant deprived of maternal company, she did what many such infants do, and became attached to a "security blanket," or towel, that she took everywhere with her. From time to time, as the towel got dirty, it was taken out of her cage and replaced by a clean one. As time went on, Gabrielle clutched the cloth less persistently, but she seemed to want to keep it, nevertheless, so Dr. Rumbaugh let her go on with it. One day, removing a dirty towel from where she had left it draped over a perching bar, he noticed that it was sopping wet. This seemed a trifle odd, but he didn't give it much thought until the next time he saw her towel draped in exactly the same place. Again it was very wet. The phenomenon was repeated every morning, until he decided to watch developments.

At this point Dr. Rumbaugh broke off and said, "I think I'd better show you what we discovered. I took a moving picture of it, because I was sure nobody would believe my unsupported word."

A projector was rolled in, and the picture thrown on one of the office's whitewashed walls. There we watched the cinematic Gabrielle, now a well-grown animal aged five years, busy at a task in her cage. She was carrying, or dragging, a towel, and as we watched, she leaped nimbly up to a sort of spigot or faucet that jutted out from a corner, not far from the crossbar on which she perched.

"That's a drinking fountain," explained Dr. Rumbaugh. "Water runs as long as you press the button, and stops when you release it."

Gabrielle held the towel under the spout with one hand and pressed the button with the other, holding it down.

"She's soaking it until it drips," said Gabrielle's owner.

I was already incredulous. I have known quite a few gibbons in my time, but I can't recollect any of them showing such efficiency and purpose. My incredulity mounted when Gabrielle took the towel, now as wet as a towel can be, and draped it over the crossbar. The camera moved downwards to where the water was dripping into a kind of hole or depression in the concrete beneath.

"She lets it drip a while," commented Dr. Rumbaugh. "And now look."

Gabrielle had descended and was lapping from the hole, now filled with water.

I said, "I don't believe it."

"Well, there it is," said Dr. Rumbaugh. "It's always said that gibbons are incapable of using tools, but here is Gabrielle using her own tool in a very elaborate way. Another thing—she drank straight out of the little pool. As you know, gibbons almost never lap: they put their fingers into the liquid and then suck it off the hairs. But that's of secondary importance compared with the main fact that she has used a cloth as a means for transporting water. Also, she has come to use a rope to facilitate her swinging." We stared at each other.

"So I really think we need a new set of tools," said Dr. Rumbaugh, and stowed his film away.

Since my visit, I have heard that members of the three other anthropoid species (assuming that gibbons belong in that class) are now receiving the benefits of civilization through a highly modern medium. Dr. Bourne was worried by the fact that these intelligent animals tend to mope in captivity, especially when they are kept in isolation for various medical tests, as they must sometimes be at Yerkes. Without stimulation, the apes' health suffers. Various symptoms of this boredom are summed up in the term "cage fatigue." It occurred to Dr. Bourne that television might provide the cure for cage fatigue, and he said so in a newspaper interview. A public relations man for Sylvania—a firm that makes TV sets—picked up the item, and Sylvania offered ten sets to the center, which offer Dr. Bourne was happy to accept. The experiment has been a complete success. According to a local newspaper, one of the chimps, Dobbs, who had been the pet of a Santa Barbara woman before he came to Atlanta, was already a TV addict before he arrived. He shows a marked predeliction for Westerns.

"Dobbs can even tell the bad guys from the good guys," said Dr. Bourne to the reporter. "He associates the bad guy with violence, and sometimes he starts jumping up and down even before the heavy says anything rough." Moreover, once when he had watched a non-Western picture in which two chimpanzee actors tore up the inside of a house, Dobbs did the same thing to his room.

The apes are not allowed unlimited viewing. On most days they are given programs for only eight hours, from 9 A.M. to 5 P.M., and then the goggle boxes are turned off. Even so, they are much happier nowadays. As Dr. Bourne said, "We feel that TV is a real contribution here."

Dr. Walter A. Pieper, another psychobiologist, was busy on a project that interests the Department of Justice very much—the effects of marijuana on smokers. "It's pretty relevant research, these days," he said, "but we're a long way from getting complete answers. The chimpanzee offers an excellent model, of course, but we ran into a snag right away: it seems that chimps don't inhale. Yes, I know they can be taught to smoke cigarettes and cigars, but for the most part, they aren't really inhaling. I'm trying to make this animal drag longer and longer."

We were back on the top floor, watching a large chimpanzee who had a sort of wheel attached to his cage. Around the edge of the wheel was a circle of little holes, some of which held cigarettes. The whole thing looked like an oversize pencil-sharpener.

"What we want is for the smoke to come into contact with the respiratory mucosa," said Dr. Pieper. (A New York doctor later gave as his opinion that it wasn't necessary to inhale marijuana to get an effect; if it came into contact with mucosa in the mouth, that did it.) "We have been trying to evaluate the animal's efficiency when he's smoked," continued Dr. Pieper, "giving him these cigarettes in rotation: one day, he has regular cigarettes, the next day regular cigarettes plus marijuana distillate, and so forth. During one session he'll have six cigarettes. We've had one or two results that might mean something: for instance, on marijuana his efficiency ratio dropped, and next day on plain tobacco it came up again. Sometimes on marijuana the ratio drops from 85 to 4 percent which is pretty striking—if it means anything, but we can't be sure as yet. Incidentally, the loss of efficiency is a transient effect. But we don't know what it all means yet."

He rubbed his forehead. "Of course, this work isn't focused exclusively on marijuana. If all the marijuana in the world were suddenly to disappear, we could still use this method for studying emphysema and other lung disorders."

He brooded for a minute, then brightened up. "We've got other problems, too," he said, "for instance, alcoholism. It's better understood than marijuana, but there is still a lot to find out, and for this we depend on Charlie. Charlie is five and a half months old now, and gets alcohol in each of his daily bottles of milk. He was drinking the equivalent of a quart of beverage alcohol a day, until he contracted pneumonia and we had to take him off it. He's recovering, by the way. As far as we could tell, he had typical withdrawal symptoms: hyperreflexia, spastic rigidity, photophobia, sweaty palms and feet, increased responsivity to tactile stimulation—all good clinical withdrawal symptoms. We're also following the progress of this case with liver studies—electron microscopic examination from a needle biopsy. The liver's showing the beginning of changes, similar to what you'd expect of a human alcoholic's liver. This work's funded in part by the National Center for the Prevention and Control of Alcoholism, which is part of the National Institute of Mental Health, but the center here provides the resource—namely, Charlie. Charlie's blood by evening shows that for every 100 milliliters of blood he has 200 to 300 milligrams of alcohol. Most states say that 150 milligrams per 100 milliliters makes you too drunk to drive, you know. At his drunkest, Charlie just sleeps: we'd say he's passed out. It's almost impossible to rouse him. Certainly he's well above the point where he's too drunk to drive.

"However, we don't want to depend on him alone, and we're beginning observations on another chimpanzee—a female. She'll be put on the same schedule ultimately, but she's still too small. You may know that there's a growing body of evidence that to some extent at least the condition of alcoholism *may* be very closely related to biochemical changes in the central nervous system. It's a very tentative theory so far, but it's respected: these changes, even if they're not the total cause of the condition, are a very important part of the origin of this ailment. Alcohol's a pharmacologically active compound, and dependence on it may be established in a similar fashion to dependency on opiates. One way to test this idea is to see what happens to closely related species like—well, like Charlie.

"By the way, we made an unfortunate choice for a control animal when we picked Jama. You know what a control is, don't you?"

"Well," I began cautiously, "it would be a control animal, wouldn't it? One that you don't do anything to—for purposes of comparison." As he nodded assent, I asked with more assurance, "Who's Jama?"

"Who's Jama?" he repeated, scandalized. "One of the most interesting chimps we've ever had at Yerkes, that's who. Or that *was* who: she's dead now. She was born here, in July 1968, and as I say, I planned to use her as a normal control animal, but pretty soon it became apparent that physically she was developing very slowly, just about half as fast as she should have. We didn't have any preconceived notions as to what was the matter with her, but obviously something was. For example, when she was seven months old she couldn't sit up or move around, which is not at all normal.

"Well, in March 1969, one of our technicians in McClure's laboratory—he's our veterinary pathologist—was making routine chromosome studies of her blood cells, and she noticed something unusual about them. McClure looked, and spotted an extra chromosome, which is to say that Jama had forty-nine chromosomes instead of the normal forty-eight. One of her doubled chromosomes, in other words, wasn't really doubled; it was triplicated, and that's the same kind of structure you find in human mongoloids. There were some differences as to *which* chromosome was triplicated: in a human it would have been the twenty-first, whereas in Jama it was the twenty-second or twenty-third, but as the great apes have twenty-four pairs and humans have only twenty-three, that might account for it.

"Assuming that Jama was a true mongoloid, it's the first time anybody has spotted a mongoloid great ape. Of course, that doesn't mean there haven't been others, only that nobody has ever recognized one before. That she died young is no wonder; the postmortem showed a lot of things wrong with her, but she had a nice personality—everybody who worked with her said she was nice, and mongoloids *are* usually gentle. It's true that she didn't like to be held, where human mongoloids are usually cuddly. But she had a nice personality."

"Jama?" said Dr. Harold M. McClure. "Yes, she died at seventeen months—she had three major cardiac defects which played an important role in her death. In addition, there was a weakened diaphragm with partial herniation and one lobe of the liver extending into the lower thoracic cavity. Both the parents are chromosomally normal. The mother had one previous pregnancy which was terminated by a spontaneous abortion, and since Jama's birth she's had another infant which seems chromosomally normal, but had a low birth weight. Jama's parents were both young—the mother was fifteen and the father twenty-two years old.

"Another research project, started six months ago, entails raising infant chimpanzees and monkeys on unpasteurized milk from cows with leukemia. This milk has been shown to contain viruslike particles, although we do not know whether this is a leukemia virus or not. It has been recently shown that leukemia in cats is definitely caused by a virus. This disease in animals is similar to that found in humans—the biggest difference being that in dogs, cats, and cows, it is usually localized in solid tissues, especially lymphoid organs, and infrequently shows a true leukemia blood picture as seen in man.

"The significance of the viruslike particles in cow's milk—or their possible effect on humans—is not known. The study which we have under way was designed to evaluate the possible biologic significance of these particles on man's nearest relatives, the nonhuman primate. At present we have only a limited number of animals in the study, and as yet have not seen any abnormalities in their blood pictures. We make monthly blood tests and physical examinations. Other determinations will include biopsies, electron microscopic studies, cytogenetics, and so on. One half of the animals in the study will be raised on control milk diets. If we find any blood changes or other abnormalities, we will then evaluate the effects of pasteurization on the cow's milk.

"One often hears that nonhuman primates seldom have cancer. During the past few years we have seen six animals with tumors. Two of these, in chimpanzees, were benign. The four cases seen in monkeys were all malignant. Leukemia has not been reported in the chimpanzee, and is very infrequently reported in monkeys.

"We also have for observation a group of irradiated monkeys who were exposed to atomic bomb explosions in 1956 and '57 at the Nevada testing site. They've been followed ever since, and I've been working on the project ever since I came to Yerkes. We have a breeding program to see the effects, if any, on their reproductive functions. Out of nineteen births we had 40 percent stillbirths, which is very high: the usual number is 5 to 10 percent. Of course, these animals are relatively old—the species lives, more or less, for twenty years. In one adult female exposed to a heavy dose, we found two chromosomes not usually found in monkeys, but she may always have been abnormal, as the animals were not studied with reference to their chromosomes before irradiation exposure. Attempts to breed her have been unsuccessful. And—oh, yes—she has bilateral cataracts and shows accelerated graying of the hair. One member of the group that died in 1968 had two primary malignant tumors—sarcoma of the thigh and a malignant tumor of the kidney. Electron-microscopic study of the thigh tumor showed a relatively large number of viruslike particles. Unfortunately we didn't do transmission studies, and the tumor cells could not be kept alive in tissue culture; consequently, we don't know the significance of these viruslike particles. The type of thigh tumor observed in this animal has not been previously reported in primates, and I think that its occurrence in this animal is most likely related to her exposure to irradiation. Kidney tumors have been noted previously to occur spontaneously in monkeys."

Dr. Charles E. Graham, whose field is cancer and reproductive biology, comes from England, where he began his work on cervical cancer histogenesis. There he worked on mice, studying the histological changes that occur when the rodents develop cancer, and trying to determine just where their tumors originated.

"I'm just finishing that study," he said, "and to tell you the truth I'm sorry I can't work any more with rodents. You see, we have no cervical cancer model among the primates. We want to develop a model for the study of histogenesis of cervical tumor in the primate. We've had two years with the squirrel monkey, injecting carcinogen—but we had no luck; we got no lesions. Now we're trying the bush baby. Also, we're in close touch with people at a hospital nearby on the Herpes type of virus —Herpes genitalis, Type Two. There's evidence that this organism is somehow connected with the genesis of cervical cancer. Frankly we're still looking for a suitable primate—we're considering cebus monkeys at the moment, because this species is susceptible to the virus, but survives its attack. We've inoculated seven animals with the virus, and we're waiting now for lesions to show up. Naturally we can't be sure of success, but we're hoping for neoplastic lesions. Anyway, we've got an animal that takes the virus, and that's something.

"The other half of my work is studying the reproductive system of the great apes—the project is funded by the Ford Foundation, and it's to last for three years. I'm working on that with Dr. Janet MacArthur at the Massachusetts General Hospital, also Dr. John Preedy here at the school of medicine, an authority on steroid hormones. We hope to obtain additional basic knowledge necessary to develop improved methods of population control. Of course, these problems are of pressing importance just now. Our main object is to ascertain the similarities and differences between apes and humans in respect to hormonal control of the sexual cycle.

"We know there *are* differences, and an understanding of these is likely to be as illuminating as the likenesses. If we understood the basic endocrinology of apes better, we could undertake more experimental studies —in fact, we've already started. We've injected labelled radioactive steroid hormone into a peripheral artery; the hormone finally passed through the gonads. Then we opened the animal's abdomen, by means of a cannula placed in the appropriate vein, and collected the blood as it left the ovary or testis. We hope to determine how the steroid hormone is metabolized as it goes through the gonad. This type of experiment is very difficult to do with humans, as you can imagine. Apes are much better animals for this than the lesser primates. Chimps are especially useful because of the sexual swelling exhibited by the females: human females show no such external evidence of curcling. This swelling is potentially a useful guide to when ovulation occurs. Even so, we are not sure just when ovulation occurs in the chimpanzee, but we hope to find out. What is the hormonal control of sex swelling? We are trying to find out."

Professor Hal Warner describes himself as an audio engineer who was in radar and worked on the space program in 1958 before coming to Yerkes. "I worked with Dr. Robinson, before he left, on impedance exploration of the brain," he said. "We find the area in the brain we want to examine, determine the local co-ordinates of the structure, then insert an electrode. Through this impedance technique, photographically, we produce, in effect, impedance-stained brain sections where white matter, a high impedance tissue, photographs light and gray matter, a low impedance tissue, photographs darker.

"We have found that live and dead brains do not differ much in their impedance profile, although the absolute impedance itself may vary. What is significant is the profile. It is possible to scan the whole brain, section by section, in much the same way that you would scan a scene for television."

Much of Professor Warner's work was explained graphically by designs and charts. One, labelled "Investigation of Intracerebral Hemorrhage," could hardly be misunderstood even by a beginner, but he willingly explained it nevertheless.

"When an artery in the brain ruptures, the brain suffers damage of several kinds—mechanical, circulatory, chemical. You see the brain contains glial cells called astrocytes, which, as some believe, like tiny placentae, take nourishment from the blood, feed the nerve cells, and remove waste. These astrocytes are believed by some investigators to form the blood-brain barrier. Since normally blood is never in contact directly with the brain tissue, perhaps some of the profound paralysis resulting from a cerebral hemorrhage is due to this intimate contact. For instance, we know serotonin can cause spasms in blood vessels. Perhaps other constituents in the extravasated blood can cause embarrassment to neurons as well as vessels. We also wonder about possible electrical effects."

"Do we?" I murmured. My mind was full of bewildering images: I seemed to be standing inside a giant brain, something like a boiler room where a very large pipe had just exploded. Hunks of molten metal were lying around, and boiling water was gushing out of a jagged hole.

Professor Warner continued firmly, "There is also the possibility of electrical damage. Could there be some ionic effect due to this blood contact, which may embarrass the propagation of the action potential?

"In order to investigate separately these damage factors, it was necessary to create an artificial bleed in an animal. Silicone oil is a biologically inert substance, so I inserted a stainless-steel cannula into the brain's internal capsule and fixed it to the skull with acrylic cement. Then I injected a measured number of cc of sterile silicon oil at normal blood pressure. . . ."

"One thing I'd like to know more about," I said to Dr. Bourne just before taking leave, "is how you used chimpanzees as blood washers. Wasn't there something about two small girls?"

He said, "Oh, yes. Two little girls—one three years old and the other five. The cases didn't occur at the same time, but they had a good deal in common: each had liver damage; each child was in a coma when she arrived; and each was linked up by rubber tubes with the chimp's blood system—the chimpanzee had O-type blood. The first child was linked up for six hours, and the other for sixteen. In the first case the blood chemistry dropped to normal after the six hours, but the child never came around: there'd been too much brain damage, and she just died. However, it was demonstrated that a chimpanzee kidney actually *would* work as a kind of kidney machine, in cleaning the blood. In the second case the little girl woke up and could remember her name and telephone number. She could even sing a little song she knew. She lived three weeks, but her bone marrow was damaged and her blood wouldn't clot. However, the chimpanzee had indubitably pulled her out of the coma. The fact is, she'd been brought in too late, as a last resort; otherwise, I think, it would have been a successful operation. The chimp got the virus, of course—infectious hepatitis—but didn't get sick: it survived very well. We quarantined it for two months, and then assumed it was free of the virus."

Not long after visiting Yerkes I learned more of this kind of cross-circulation from reading abstracts of talks that had been given at a conference on experimental medicine and surgery in primates, held in New York in 1969 and sponsored by the New York University School of Medicine. Five doctors had collaborated on an experiment closely related to this work—Edward I. Goldsmith, J. Moor-Jankowski, Alexander S. Wiener, F. H. Allen, and Robert Hirsch. "Temporary cross-circulation between human patients in liver failure and nonhuman primate partners may be clinically valuable," they wrote, "but the methods of preparation of donor animals and the immunological consequences remain to be defined." They had therefore worked out a program for choosing "appropriate animals" and preparing them as safely, simply, and economically as possible for the process. Having ascertained which animals had blood close enough to the human types for the purpose, the doctors had taken two species, rhesus and baboon, and "exchange transfused" them—i.e., replaced their own blood with human blood. The procedure apparently requires the most careful attention "to preoperative typing and cross-matching and body temperature maintenance, blood pressure monitoring and calcium replacement during exchange," but with all this taken care of, the animals can survive on virtually 100 percent human red blood cells for four to five days, and during the first two days of this period they might serve as cross-circulation partners for human patients.

In fact, as another report at the same conference testified, the experiment was actually carried out at the Rochester Medical Center, where cross-circulation was arranged between a woman of sixty-six, who was in deep coma from an advanced stage of hepatitis, and a baboon prepared according to the technique described above. The baboon was typed and matched with the patient and a cross-circulation lasting five hours was instituted. During the first three hours of this period, "the patient's level of consciousness improved slightly, although she did not become conscious. . . . There was improvement in the results of all tests as the cross-circulation proceeded," and after it was discontinued, she actually woke up for a little while and spoke a few words. But she soon lapsed back into coma, and six days later she died. The autopsy showed that her liver was completely necrotic—that is, the cells were dead.

Another baboon was used in similar fashion, but with a happier outcome, for cross-circulation with a twenty-one-year-old woman waiting for a liver transplant. She had gone into a deep coma which did not respond to the usual methods of treatment, and according to the report, "Cross-circulation was carried out for eight hours with dramatic relief of the patient's coma."

I made the acquaintance of two of the originators of the preparation method, Dr. Jan Moor-Jankowski of the New York University School of Medicine and Dr. Edward Goldsmith of Cornell University Medical College, at the place where they spend much of their working time—the New York University Hospital's primate laboratory in Sterling Forest, near Tuxedo. To my eye, accustomed by this time to large, rambling primate center buildings, the administration offices of the laboratory seemed charmingly small and simple as well as pretty. It harmonized well with the surrounding rocks and trees. Dr. Moor-Jankowski seemed pleased when I said as much. "We have spent as little as possible on building," he said. "This is the only permanent part of the place: all the laboratory sections and animal shelters are in trailers which can be moved around when we find it necessary."

He led the way to a clearing in which stood the prefab laboratory—large trailers like railway coaches, with a shining metal finish. Inside these coaches, once one had learned to discount the hollow feeling of one's footsteps, it was easy to forget the temporary nature of the housing because the equipment was so complete. There was a trailer like an office —well, it *was* an office—and behind it were laboratories, all carefully kept antiseptic. In a separate cluster of trailers were the accommodations for chimpanzees and other primates.

The laboratory section had everything, even a movie projector, where I was shown pictures of research on filaria—threadlike nematode worms that get into the bloodstream from infected water. Filaria are a scourge in the warm regions of both hemispheres. Like men, baboons (as well as other mammals) are vulnerable to them, and by using baboons, a means has been contrived to filter these parasites out of the blood by mechanical means.

Dr. Moor-Jankowski then led me into another trailer where I watched a chimpanzee being prepared for a routine inoculation. She was a young animal of four or five, but she knew all about the squeeze-cage (it is a contraption for holding the occupant motionless) from earlier experience, and she wanted no part of it. She screamed with rage and fear until one of the attendants managed to shoot into her a sedative much used in animal laboratories, called by its trade name, Sernylan.

"In a few minutes she'll be all right," said Dr. Moor-Jankowski. It was true: very soon she calmed down and looked quite happy. If it didn't sound anthropomorphic I would say she was dreamily euphoric.

The same drug, but in a stronger dose, was administered to a large adult male baboon whose sperm was wanted for experiments in artificial insemination. He quickly went to sleep, and was taken from his cage perfectly safely.

"Artificial insemination with primates isn't as easy as it is with cattle," one of the doctors told me. "With subhuman primates, anyway, almost immediately after ejaculation the stuff hardens and turns into a sort of tough jelly. In natural copulation it forms a plug and keeps the sperm safe inside the female: I suppose that's how it developed. Otherwise arboreal animals would lose the seed through cavorting around in trees. But it makes things difficult for us here. We're trying if we can quick-freeze it and yet keep the organisms alive, but the method hasn't proved satisfactory so far, and after all, if you've got to use the sperm immediately to get any good out of it, you defeat the whole object of the exercise."

Dr. Moor-Jankowski's chief interests, as one might judge from the reports from the conference, lie in the field of serology. A few days later he arranged a visit for me to the office of Dr. Wiener, who is professor of forensic medicine at the New York University School of Medicine and city serologist in the office of the Chief Medical Examiner. Both these departments were represented, in a way, in his laboratory. At least, the place was full of test-tube racks and mysterious computerlike machines, while in the corner was a broken paving stone with a large brown stain spilled on it. Dr. Wiener saw me looking at that in a puzzled way, and explained, "They've sent this to me to find out if that stain is blood: someone's coming up for trial in a murder case, and this stone is part of the evidence. It's blood all right."

He talked to me of the similarities and dissimilarities of human and subprimate blood, of the various antibodies, and of titers with which these are measured. In his enthusiasm he went so thoroughly into the subject that once or twice I saw a female assistant glance at him anxiously and then look at her watch. After the session, happening to meet me alone in the hall for a minute, she said drily. "Well, how did you enjoy getting the benefit of twenty years all in one afternoon?"

Over coffee, Dr. Wiener grew reminiscent about his early days when he had hesitated as to a career. "I liked mathematics, too," he said, "and for a time I thought of specializing in that field. Even today when I want recreation I turn to mathematics."

Once he had settled for serology, he said, he found it hard to get hold of samples. "Today, with the laboratory in Sterling Forest and good communications we have with other places all over the country, we aren't faced with the same difficulties," he continued, "but in those days—it wasn't so long ago, either—it was a serious matter. Once when I wanted a sample of chimpanzee blood, I applied for permission to get some from a chimp in the Central Park zoo. Today the administration of the zoo is handled by people who know what they're doing, but then the man in charge had been given his job as a political plum, and he didn't know anything about scientific research. Still, I had to apply to him.

"I told him what I wanted and assured him that it wouldn't hurt the animal, and so forth. He said it would be all right, but that I had to put down five hundred dollars in advance as a sort of insurance, in case the animal suffered damage or something: also, I'd have to pay the keepers who helped me. Well, I handed over the five hundred, and that was the last I ever saw of it—not that I'd expected anything else. Then I fixed things up with the keepers, two of them. They told me to be there after closing hours so the public wouldn't see us, so it was getting dark when I met them in the monkey house.

"The chimp cage was down at the far end of the place, past all the other monkeys. It was a simple operation—one of the men held up a banana outside the cage bars, and when the chimp reached out to get it, they both grabbed his arm and held on tight while I hurried to take the sample. Then they let go, handed him his banana, shook hands with me and collected their pay, and left by the side door while I started down to the front one.

"I had two test tubes of chimp blood in my vest pocket, and I was feeling pretty pleased with myself, when all of a sudden something hit me in the back of the neck with quite a wallop—something wet. The chimp had spit at me, all that way, and in the dark at that. Remarkable aim he had."

5

The Power of the Inductive Method

Dr. Harry F. Harlow, head of both the Wisconsin Regional Primate Research Center and the Department of Psychology Primate Laboratory at the state university in Madison, Wisconsin, once told a friend how his particular branch of research happened to get started. It was before the center was formed, and Harlow's domain was limited to the laboratory. That day he was showing the place to a distinguished visitor, the English psychiatrist Dr. John Bowlby. Standing between two rows of cages in each of which a rhesus monkey sat alone, one after another, cage after cage, the two men spoke of Pavlov, who was experimenting with neurosis-like behaviors in various nonhuman animals.

"I wish I could get hold of a few neurotic monkeys," said Harlow fervently. "If only one could find them!"

Dr. Bowlby bent on him a mildly astonished gaze. "But you've got dozens of them right here," he said, with a gesture that took in the whole room. "How many do you want?"

"He was right, of course," said Harry Harlow in retrospect, "but we simply hadn't thought of our monkeys' behavior in that way: we were used to it. We were aware that if a rhesus is kept in isolation long enough it develops all sorts of psychopathological behaviors—thumb-sucking, or self-hugging, or just staring into the distance—but we needed a psychiatrist to point out the obvious fact that we'd been raising monkeys for years with just the sort of abnormalities we needed to produce syndromes analogous to human disorders."

Thus when, in 1961, the NIH tapped Madison as one of the seven locations for the primate centers they were planning, a foundation was already there. To the laboratory staff the center plans were more than welcome: they had had to fight hard for what they held, and had long been haunted by a sense of insecurity. When Harlow came to Wisconsin in 1930, he discovered very quickly that as a member of the Psychology Department he was, to say the least, underprivileged. On asking his chief to show him to the animal laboratory, he was told, "We tore it down last summer."

I remember that laboratory; I used to fool around in it as a child—a wooden bungalow structure of hexagonal shape, in which a few guinea pigs and rabbits moped in makeshift enclosures and nobody was there to tell me not to pet or feed them. For all its shortcomings, however, its disappearance must have been disconcerting to the newly arrived Harry Harlow. For lack of anywhere else, he went to work at the local zoo in Vilas Park: fortunately, the zoo had a fairly varied collection of primates, and primates were what he wanted to work with.

In a talk he gave during a meeting in the sixties, he cast his mind back to those pioneer days in the park, to an orang-utan named Jiggs. "The nicest and sweetest orang-utan that had ever lived at any zoo for fifteen years," he said. "We gave him two oak blocks, one with a square hole and one with a round hole, and a square plunger and a round plunger. He learned to put the round plunger in the round hole and the round plunger in the square hole, and he learned to put the square plunger in the square hole, but he never learned to put the square plunger in the round hole. He worked incessantly on this unsolvable problem for six weeks and then died of perforated ulcers, but at least he died demonstrating a level of intellectual curiosity greater than that of many University of Wisconsin students."

Two years of this sort of experience, however, proved that working in a zoo had many disadvantages. Harlow and his crew really had to have a better place, and they found it in an old two-story building that had formerly belonged to the Forestry Department. The psychologists moved in and began remodelling, an activity which continued for twenty years thereafter. They rebuilt, added space, fought grimly for the necessary equipment, and—almost incidentally—learned a good deal about monkey behavior. Out of all this came three major research projects: the development of measures of subhuman primate learning, the effects of cortical lesions on these learned behaviors, and the analysis of primate motives.

In 1954 the laboratory was moved wholesale to a remodelled cheese factory—Wisconsin's agricultural college was comparatively privileged, and didn't need it any more—and at last, in 1959, it actually gained the dignity of a proper name of its own, the Department of Psychology Primate Laboratory. Legends began to accrue around it.

"One of the brightest monkeys we ever had was number 130," said Dr. Harlow. "She was living in one of our combined indoor-outdoor cages, and there were always two doors between her and freedom. This did not bother number 130 in the least because she outthought the graduate students on the average of once a week. She knew all about people and knew that they were careless, indifferent, shiftless, and stupid. She would bide her time, and at the chosen moment the graduate student would find himself in the cage and find 130 outside the cage. She spent more than half of her last summer in the upper branches of a large elm tree, and I think that the only times she let herself be caught was when she read that rain was predicted. However, her peregrinations in no way interfered with our testing program because she could solve as many problems in two days as any other monkey could in seven."

One of Harlow's assistants at this time was a coed named Audrey. "We had six baby monkeys that we had separated from their mothers at birth," he says, "and we kept each monkey isolated in a separate cage. We wanted to keep them isolated for two years and then study their initial social responses. Audrey was taking care of them on the night shift. Late one evening I went over to the lab, opened the door to the baby room, and there was Audrey sitting in the center of the floor playing with all six monkeys. They all had the same I.Q. and they were having a wonderful time.

"She didn't give me a chance to say a thing. She arose to her full five feet two inches of enraged femininity and she said, 'Dr. Harlow, I'm a student in secondary education, and I have been trained, and I know that it is improper and immoral to blight the normal social development of children. I am right and you are wrong.'

"We didn't know that we had been paying her to serve as a baby-sitter," adds Dr. Harlow. "One year's time and five thousand dollars forever lost. I was angry, but I also had some respect for little Audrey. I felt that perhaps we had something in common. One must have faith in what one does. However, I can assure you that Audrey and those monkeys never played together again.

"Four years have passed, and Audrey has twice been pregnant, which is more than one can say for any of the monkeys. Last week I observed these monkeys in their cages, sitting staring into space, sucking their thumbs, engaging in unending stereotyped movements. It came to me as a complete shock; the monkeys were abnormal, and their social lives *had* been blighted. Audrey was right—there is apparently something to this secondary-education business after all. However, it wasn't necessary for Audrey to tell us. We found it all out through research—but it took us four years longer. This illustrates the power of the inductive method. If you work hard enough and long enough, you will eventually learn some of the things that other people already know."

Since its dedication in April 1964, the Wisconsin Primate Center has taken on that indescribable patina of a university building that belongs on the campus just as certainly as do its neighbors. A brief description furnished by the NIH told me that its main research orientation is, naturally enough, behavioral and physical development, but other and related areas are "brain function, mental retardation, biochemical mechanisms of learning, abnormal behavior, nutrition, pregnancy radiation, cardiovascular disorders and amino acid metabolism." Between them the center and laboratory, along with a "holding facility" at Vilas Park Zoo, maintain eleven hundred animals, most of them rhesus.

By good fortune I had a young friend, Connie, who was taking her Ph.D. at Wisconsin just then in behavioral psychology, under the guidance of Associate Professor Gene P. Sackett of the Department of Psychology Primate Laboratory. Sackett has an impressive lot of publications to his credit: "Some Effects of Social and Sensory Deprivation During Rearing on Behavioral Development of Monkeys" is one example. Published in the *Revista Interamericana de Psicologia* in March 1967, it says among other things that according to Dr. Harlow's findings, a year of isolation is particularly devastating for the animal's socialization; also, that monkey infants who have never been near a real monkey respond to pictures of monkeys, even reacting to threatening poses in those pictured animals.

Connie took me to see a number of infant monkeys—baby rhesus, *I* would have called them—troublingly appealing with their small, wrinkled faces. One of them, as if to get a better look at me, clung upside down to the bars of her cage.

"I'm comparing their rates of development," Connie explained. "All sorts of influences moderate this. For instance, I have a monkey infant living with its mother. Well, I compare its behavior and learning ability, and so on, with another infant deprived of maternal support. We are also bringing up certain monkeys that just live in their cages without anything to do—no stimulation—and we compare their accomplishment in tests with that of other monkeys that live in an enriched environment."

"What does that mean, an enriched environment?"

Connie gave me the sort of look she usually reserves for her mother.

"More space and plenty of toys," she said. "Offhand, before we've evaluated our findings, it looks as if monkeys reared in these privileged surroundings prove to be more aggressive and quicker to learn than the deprived ones. Within peer groups, they tend to dominate."

A minuscule leader in the making deserted her mother and flopped across the cage to threaten me, opening her tiny mouth to show almost invisible teeth.

"Surely you knew that before you did the experiment, didn't you?" I asked. "I mean, of *course,* privileged monkeys would be more bossy."

"You can't leap to conclusions," said Connie severely. "In this work the great danger is anthropomorphism."

I found Dr. George P. Kerr, from the Department of Pediatrics at the Medical Center of Wisconsin University, explaining a piece of research to three young visitors.

"We're investigating malnutrition and its effect, if any, on the brain," he announced. "For some years I've been making observations of human children who've been undernourished, and now we're comparing them with infrahuman primates with the same background. You see, some people believe that certain children who don't grow as rapidly as they should are victims of what's called 'deprivational dwarfism,' because they have lacked environmental stimulation.

"The question we posed ourselves was whether this slow growth might not be due, instead, to simple malnutrition. We tried raising infant monkeys under various conditions that might have victimized such a child—less food, for example, or enough food but an otherwise deprived environment. Now we know what a child's normal growth rate is: it's very abrupt in early life—that is, the child grows very fast—and then, after a while, the rate slows down, and on the graph the curve begins to flatten out. We know that the same condition, more or less, obtains with infant monkeys. Therefore we have raised a number of animals in isolation but with the normal amount of nourishment: these animals developed behavioral abnormalities, but their growth rates were comparable with those of normal animals.

"The result suggests that social stimulation is not a prerequisite for physical growth, and it indicates that deprivational dwarfism is very probably caused by inadequate nutrition rather than lack of social stimulation. When we look again at undersized human children, we are led to conclude that they have simply not been fed enough. Mind you, this may still depend, indirectly, on social conditions. A child that is unhappy, even when it is hungry, may very well lose its appetite. When social conditions improve and the child is happier, it eats a lot."

He showed us on a projection screen the graph of the growth of a child who was half starved up to the age of two but was properly fed from then on.

"Notice that the weight catches up. But height and bone age do not," he said. [Bone age is the state of development of the bone in relation to that of a normal individual.] "It's a general rule that primates stop growing at puberty, but if puberty is delayed growth continues. This little girl's growth is only retarded, and will catch up in time. It appears that the time of puberty depends on bone age. As for the rest, there is no doubt that her learning capacity has been somewhat affected by malnutrition, or—this is the question—is it a simple cause-and-effect pattern or merely coincidental? We can't draw conclusions until we know more, but there can be no question that there is at least an indirect relationship between nutrition and retardation."

Next he talked of PKU, or phenylketonuria, a congenital faulty metabolism of the amino acid, phenylalanine. This amine is essential for the production of proteins, but too much of it produces mental retardation. If the condition of PKU is diagnosed in a human baby soon enough after birth and he is put on a special diet for the next six years, chances are that he will be normal or at least nearly normal. But when a woman with PKU becomes pregnant, the excessive amino acid in her blood gets into the unborn infant at an even more concentrated level, and the unfortunate child is born retarded. So far, nothing much can be done for such cases of inheritance and concentration. But Dr. Kerr and another Wisconsin scientist, Dr. Harry Waisman, have been able to reproduce the PKU condition in pregnant rhesus monkeys with high phenylalanine diets, and the infants after birth show deficits in learning. The condition in these monkeys can be studied intensively, and perhaps a cure can be worked out.

Dr. John W. Davenport is also working on mental retardation at Madison, with rats as well as primates. "They're easier subjects," he said as we watched, on one of those closed-circuit TV sets, a number of hypothyroid rats going through their paces. "Of course, they're a lot more limited, but we can produce whole batteries of retardate rats in a comparatively short time, look at their learning deficits, and get good leads for our primate studies on thyroid function and intelligence. Our primate studies of mental retardation, including those on hypothyroidism and malnutrition, are going even faster now because we have a computer for massive I.Q. testing of monkeys. We can work with larger intelligence batteries. We've worked out tasks that really do show significant differences between normals and hypothyroids. It's a funny thing, but all this grew out of a quite different research problem on studying a drug called TCAP which was supposed to *increase* I.Q. because of its ability to stimulate brain metabolism. But the drug turns out to be a potent antithyroid agent as well, and we have been finding permanent learning deficits from this drug, similar to those in rats and monkeys whose thyroids have been destroyed.

"One of the questions we have to answer is, What *kind* of intelligence is affected by these various agents? Even now it might be ten years before we can answer with any assurance, but with the computer we can telescope, and do in one year what used to take three.

"In our primate studies, thyroidectomy is producing some fascinating puzzles. For instance, Dr. Kerr and I have injected pregnant monkeys with radioactive iodine: the iodine carries over into the baby—enough to destroy the baby's thyroid gland—and at birth the baby shows many behavioral signs of cretinism. But the physical growth rate seems unaffected even though the textbook dogma for many years has held that thyroid hormones are essential for over-all fetal growth. If these infants are given thyroxin, the prenatal damage to the nervous system is not reversed, as one might expect, but seems to persist, though it's early days for us to be sure it does." He paused and thought. "We have an oddball situation here," he added. "A lot more is known about hypothyroidism in the human than in monkeys."

In Dr. Harlow's department of psychology, a playroom behind a large plate-glass window was occupied by four young rhesus, two of which had been kept in isolation until recently, whereas the other two had grown up in ordinary social groups. Researchers watched solemnly and took notes of their antics. The object of the study, I gathered, was to find out how long it took the isolates to recover from their stereotyped behaviors and join the gang like normal monkeys. The groups were changed from time to time, and the histories of the isolates taken into account, for the length of time they had spent in isolation had a direct effect on the length of time it took them to recover. Those that had been alone for three months, when put back with their peers—to use one of those sociological words that are taking such a prominent place in modern vocabularies—have proved to make dramatic comebacks in behavior quite suddenly; though until lately, I was told, they were believed irrecoverable.

"The social effects of such rearing experiences are now well established," according to the center's publication, *Mainly Monkeys.* "Subjects reared for the first 6 months of life in total isolation exhibit permanent deficits in exploratory play, sexual, and maternal behavior, while subjects isolated from birth for longer periods of time (12 and 18 months) are almost obliterated socially."

The article goes on to say that those reared for long periods in only partial isolation, where they can see and hear but not touch other monkeys, display some positive social behaviors, but in all situations are socially inferior to normal control monkeys. They also exhibit various special behaviors that can be broadly categorized as psychopathological. If the partial isolation is prolonged, the animal goes round and round its cage and even attacks and wounds its own body.

Mental depression on a wide scale is a phenomenon of our age, and for some years Wisconsin researchers have tried to produce genuine depression like that of human beings in monkeys, so that methods of treatment might be evolved. The difficulty has always been that the subjects, though they do exhibit symptoms analogous to our depressions, make such quick recoveries that scientists have had to conclude that their syndromes are not signs of the true malady found in humans. Recently, however, research people at Madison have discovered what might be an effective way of reproducing this mental disease. As any nurse in an old-fashioned children's hospital in England can attest, human children from six to twelve months old, if separated from their parents for long periods of time, exhibit certain well-known symptoms: dejection, stupor, lowered activity and general withdrawal, in that order.

Dr. Bowlby writes that slightly older children react in the same manner, and he divides this reaction into three phases: (a) initial protest, exhibited by increased crying, screaming, and general activity; (b) despair, and, following reunion, (c) detachment, where the children, reunited with their parents, appear indifferent or even hostile. (But Bowlby has recently said that he is now not sure of phase three: it is not an invariable part of the reaction.)

Baby monkeys in similar circumstances behave in much the same way except that the reaction is not so prolonged, and there is no detachment behavior upon reunion. "It may be," say the researchers in an article in *Mainly Monkeys,* "that the technique of separation from an attachment object is, by itself, insufficient to produce such desired behavioral effects. We view this finding with little surprise for we firmly believe that any behavioral syndrome as severe and complex as depression is unlikely to be mediated by any single variable." Depression or depression-like behaviors can only be produced by a combination of depressing circumstances: merely one such circumstance is not enough. To this end the researchers have worked out a combination of trials for the subjects: social isolation, *plus* separation from monkeys and things that they like, *plus* partial restriction of movement, *plus* scary happenings, *plus* alteration of day-night cycles—a change that seems to worry them.

"We wanted to produce a *stable* depression," explained Stephen Suomi, who is working on this experiment, "so that we would have time to work with one depressed subject before it recovered by itself. Thanks to an invention of Harlow's, called the vertical chamber—in fact, it's more or less a pit—we can achieve total isolation combined with restriction of movement. First the monkey is removed from its mother and isolated in a vertical chamber. It is then placed for a while in a peer group. Then it's isolated again. Well, we think we have a real depression at last; at least we know that beyond thirty days of treatment of this sort the change lasts for at least six months. This gives us a real chance, now, to find out how to treat it—to reverse the depression. Later we plan to use drugs to alter the biogenic amine metabolism, because some human patients do seem to exhibit altered biochemical levels of corticosteroids, and so on. We mustn't be overconfident: we have to be very sure that this is a true mental depression we've got, and that spontaneous recovery alone does not occur—as it has done in the past with the incomplete depressions we've produced. Then, when we are quite sure, there'll be many more experiments: How can we reverse it? Antidepressant drugs, perhaps? Another question to be answered is, How susceptible are people to all these depressant factors? Can we immunize them in advance? Such problems are still in the future, but I don't think it will be very long before we can consider them seriously."

"We've got four to five hundred breeders at the center," said Dr. Stephen G. Eisele, whose province is the breeding program. "They're in various different categories: we have 170 pregnant ones right now, and we try to space out the births. Some go to Dr. Harry Waisman for drug research, on the effects of certain artificial sweeteners, for instance, and some are needed by Dr. R. C. Wolf for physiological research—about ten to fifteen, depending on what he's doing—many go into mental retardation or depression studies, and others are strictly for fetal research. It's wonderful what they can do in that field. Nowadays you can take a fetus out, inject it, replace it, and it will go full term and be born successfully. We can study ovarian blood now. Also, we've had unusual success with artificial insemination, which isn't anything like as simple with primates as it is with domestic animals—lots of people said it couldn't be done at all. But we do it. I'd say that twenty to twenty-five of the males here have bred all our five hundred mothers.

"In the early days we bred animals only for Harlow and his mother-love studies. This was often difficult because the maternally deprived animals didn't know how to copulate, and our first problem was how to impregnate those deprived females, but now that we know a lot more about the behavioral difficulties it doesn't matter. We know that the ovum always drops fifteen to nineteen days before the menstrual period, and that occurs every twenty-eight days for seven or eight months of the year. Then, for some reason we still don't understand, an atypical pattern sets in for the remaining four months of the year.

"The female rhesus is a beautiful subject for this work because of the coloration of her sex skin: you can always tell exactly what stage she's at just by looking at the color between her legs. Here, you can see; we've drawn up a color chart. To use it, you keep in your head the color the female is today and correlate it with yesterday's color. Here's the strongest red, before it starts getting paler day by day. Two days before the color breaks down you have a 70 percent chance of ovulation, and this phase can last anywhere from five to twenty days, though typically it's five.

"Now here's an interesting contraption. In the past, Dr. Harlow raised monkeys in isolation or semi-isolation, and many of the female animals, as they matured, did not show basic sexual behaviors and consequently were very difficult to impregnate. So this restraining device was built to hold the female in proper position while the male copulated with her. We call it the rape rack. Nowadays, though, with improved techniques in artificial insemination we don't have to use it. Artificial insemination is much more efficient. One ejaculation can be used to serve as many as twenty-five females, but usually we stretch it only to ten."

I went from Dr. Eisele to Dr. Wolf, who talked on a related subject; the hormonal requirements of primates during implantation in pregnancy—when the fertilized ovum attaches itself to the inside of the uterus. In other animals, such as the rat, if ovarian estrogen is not present, the pregnancy is interrupted, but this does not seem to be true of the rhesus monkey. Dr. Wolf and two colleagues removed the ovaries from eleven female monkeys that had been mated, between the second and sixth day after ovulation, then injected them daily with progesterone—but not estrogen—until the day of parturition. Eight of the animals maintained pregnancy in spite of the operation, but one of these aborted soon afterwards. Then two more monkeys were mated and ovariectomized before implantation, but these were not given the progesterone: they did not maintain their pregnancy."

This work, of course, has bearing on the Pill and other methods of population control.

"Shigella is the great problem in all monkey colonies," said Dr. W. D. Houser. "It's universal." Shigella are the bacteria that cause bacillary dysentery in humans. "Everybody who works at a center gets what we call 'monkey fever' within the first few days—nothing serious, diarrhea and slight fever, probably from inhaling fecal material in the air or by some similar ingestion. Then you build up an immunity, and it doesn't bother you any more. One expects it.

"But what *was* interesting"—here his eyes lighted up—"was an infec-

Chimpanzees were rare and expensive animals in 1923 when Robert M. Yerkes began to realize his dream of establishing a breeding colony of primates for scientific purposes by spending $2000—most of his life savings—to purchase Chim, right, *and Panzee, seen here with their owner on his farm in Franklin, New Hampshire.*

Yerkes trained his animals to eat together like well-behaved children. At lunch on the farm in August 1926 are Pan and Wendy, Bill (*named for William Jennings Bryan*) *and Dwina* (*short for Darwina*).

Yerkes found that chimpanzees learned through trial and error. In this early experiment, Bill successfully fits a key into a lock, above, *then eats the food that was contained in the box,* below.

A unique New England sight was that of chimpanzees climbing and swinging the birches on the Yerkes New Hampshire farm. Below, *Yerkes' daughter, Roberta, takes Pan, Wendy, Bill, and Dwina for a stroll.*

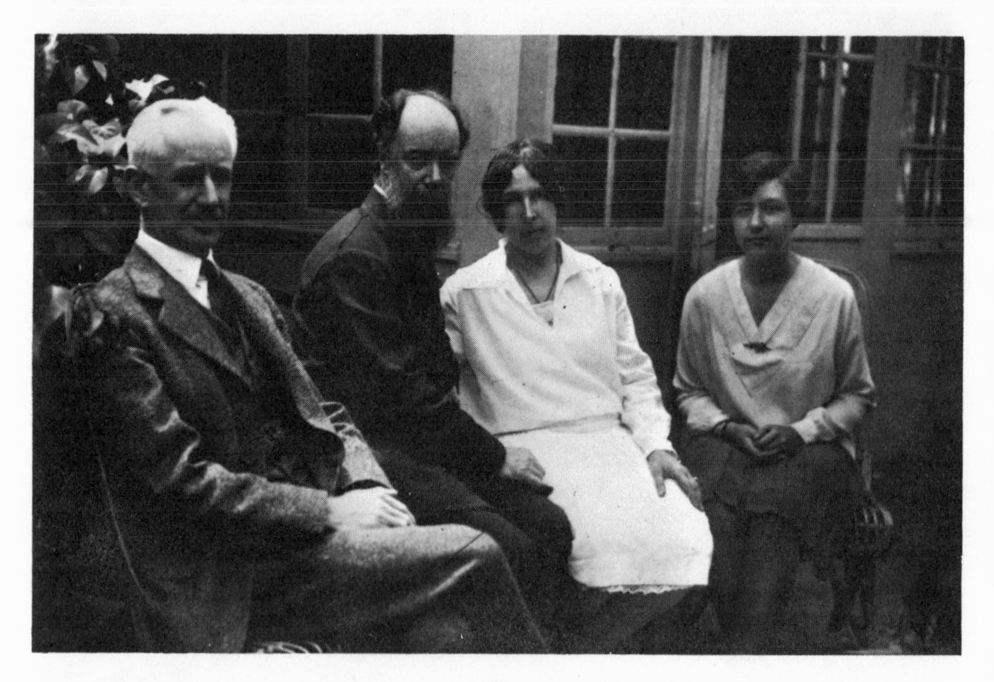

In 1929, Yerkes and his daughter Roberta visited primatologists in Russia and Africa. Above, left to right, *Yerkes, Alexander E. Kots, Nadazhda Kots—one of the first Russians to investigate chimpanzee intelligence—and Roberta, in the Kotses' garden in Moscow. Soon after returning to the United States, Yerkes moved into the new laboratory,* below, *at Orange Park, Florida.*

Pastoria, above, *the primate station and laboratory operated at Kindia, French Guinea, by the Pasteur Institute of Paris.* Below, *the cages at Quinta Palatino, the Havana estate of Madam Abreu. Yerkes' colony in Orange Park was stocked largely with chimpanzees from these two places. The Pasteur Institute gave him sixteen animals in 1930, and he received fifteen from Cuba in 1931 after Madam Abreu's death.*

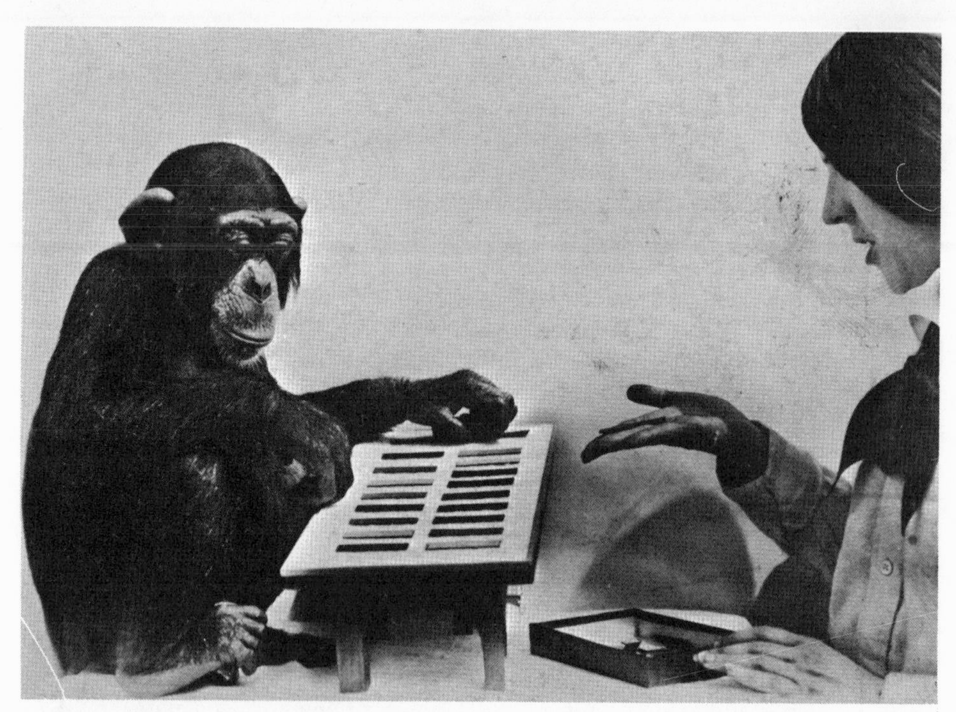

Since standard testing materials were not available in Moscow, Mrs. Kots used color samples of her own making in a series of experiments that showed chimpanzees could distinguish one hue from another. Above, *Joni matches the sample object in Mrs. Kots's hand.* Below, *Joni gives her selection to the experimenter.*

Chimpanzee using a pyramid of boxes which it constructed in order to reach suspended bananas, in an early intelligence test by Wolfgang Köhler, the Gestalt psychologist, whose work on Tenerife helped inspire Yerkes to study primates.

Peck, a 140-pound chimpanzee, smokes a cigarette as part of a study on the effects of marijuana at Emory University's Yerkes Regional Primate Research Center. The colony, founded by Robert M. Yerkes in Florida, was moved to Emory in 1956; subsequently it became one of the seven Regional Primate Research Centers supported by the National Institutes of Health. Below, *a young female chimpanzee in the nursery playroom at the Yerkes center.*

Shamba, an adolescent female gorilla, weighing 114 pounds, displays the typical gorilla walking stance as she comes to the front of her cage at the Yerkes center. She participates in learning and behavioral studies.

Dr. Harry F. Harlow, director of the Wisconsin Regional Primate Research Center, Madison, and pioneer in studies of maternal affection among rhesus monkeys. Right, *an infant rhesus monkey manages to feed from wire surrogate mother without leaving cloth surrogate.*

Two infant monkeys on the floor, who have been raised in isolation at the Wisconsin center, show symptoms of depression while their peers, reared in normal social groups, play on the ladder.

An adult rhesus monkey at the Wisconsin Regional Primate Research Center.

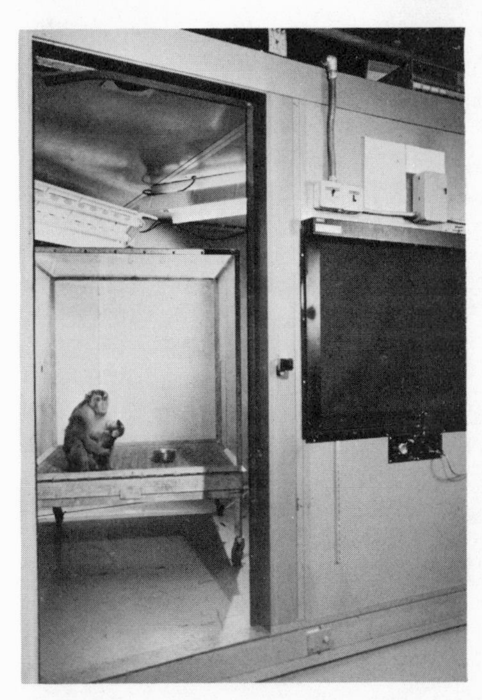

Below left, *Mother and infant pig-tailed macaques in a "deprived environment" at the University of Washington primate center.* Right, *rhesus monkey is restrained in a special chair to prevent injury after surgery at the University of Oregon primate center.*

Brunie, a pig-tailed macaque, with her own infant and one she adopted, at the Washington Regional Primate Research Center.

Animal handler Jan Holte and young rhesus monkeys in a colony at the Oregon Regional Primate Research Center, which also maintains intact a troupe of Japanese macaques (below), *just as it was captured in the wild, complete with leaders, females, bachelors, and children.*

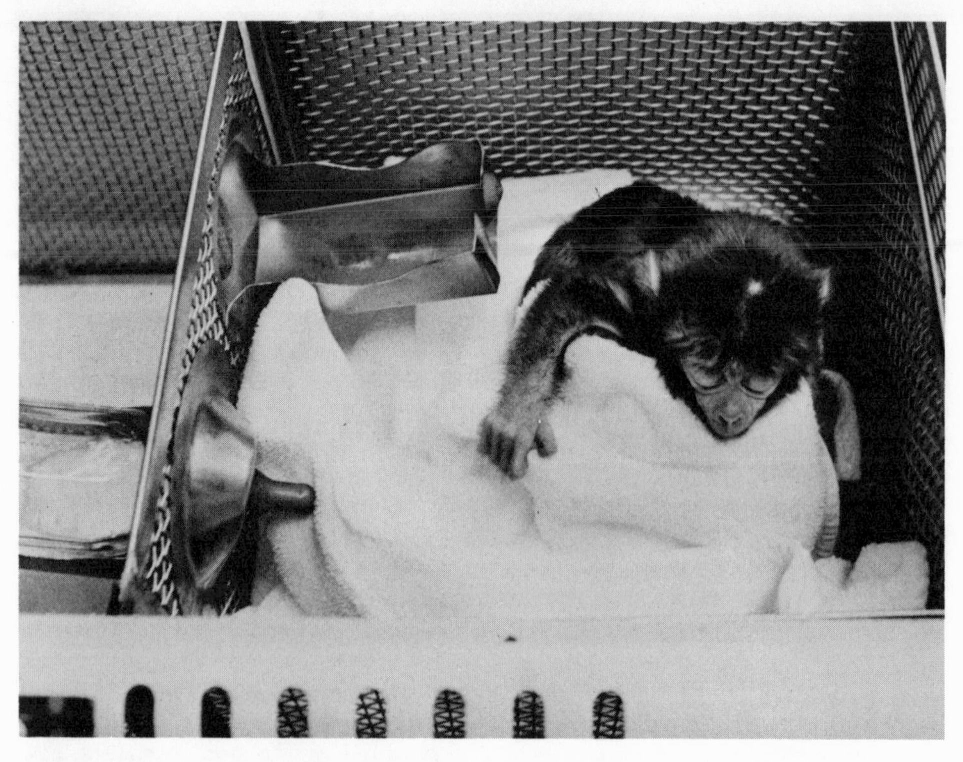

Three-day-old rhesus monkey in incubator at the Oregon center.

Left, *a tarsier, used in studies of reproductive physiology and behavior, at the Oregon center.* Right, *a ring-tailed lemur with twin infants at the Oregon center.*

A rhesus monkey awaiting a chest X ray at the Oregon center.

Corn cribs were adapted as cages for rhesus monkeys at the National Center for Primate Biology in Davis, Cal. The center's colony of about fifty Indian langurs, below, *has been used to study plague.*

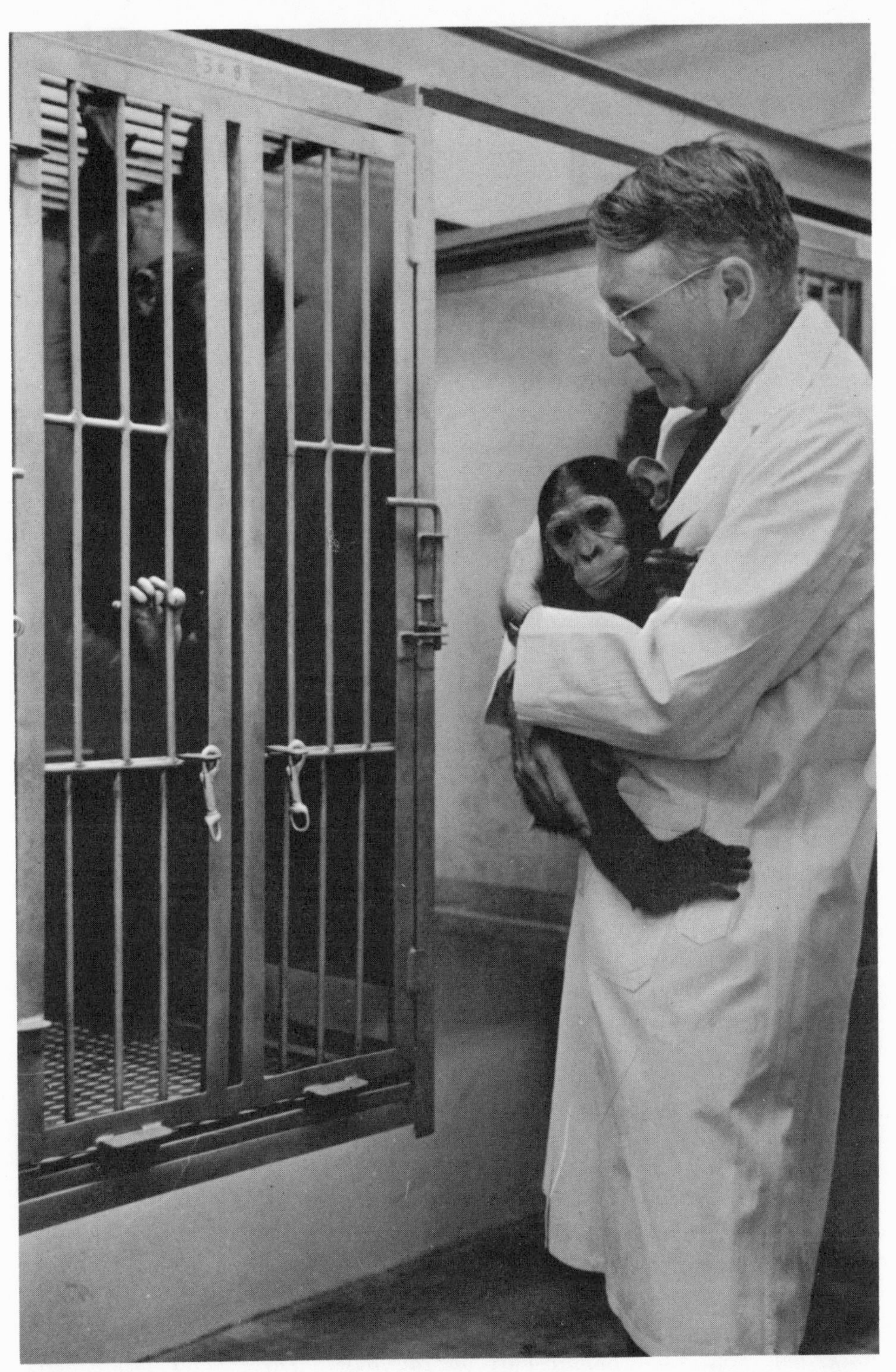

Dr. Arthur J. Riopelle, Director of the Delta Regional Primate Research Center, Covington, La., holding a young chimpanzee.

"Tandem walking" at the Delta center. Rock-Gigi followed by Libi-Bido. Chimpanzees show strong preferences for the same partners, and the same animal usually "leads," though the one in rear often does the "steering."

Belle, the leader of this group, uncovers a rubber snake with a stick, while Rock, left, *Polly, and Bandit look on cautiously. Three other chimpanzees became upset and left the scene, even before the snake became visible.*

Begging Food: Shadow eats a fresh pine cone.

Bido takes a taste.

Bido tries to take the cone, and Shadow hands it over without protest.

Copite de Nieve, or Little Snowflake, the albino gorilla in the Barcelona zoo. When Delta's Arthur J. Riopelle handed him a white ape doll, a replica of himself, Copi recognized the similarity right away.

tion of *Klebsiella pneumoniae* we had in the spring of 1969, an atypical form that hadn't been reported before, in infant monkeys. First we noticed that a few monkeys were getting very depressed, and soon we had all kinds of things—meningitis, peritonitis, cystitis, arthritis, and subdermal abscesses. I can tell you, we were busy. Three monkeys died, and the other cases needed a lot of therapy before they recovered. As far as we can make out, the infants were infected at birth, through the umbilicus. The organism resisted most antibiotics, until at last we found a couple that worked. We drained the abscesses. They're all right now. First reported case in history," he repeated proudly.

During my latest visit to Madison, I found Connie busily taking notes in a room containing two cages not side by side, but connected here and there with wires and tubes. Each held a baby monkey and a number of tiny, mysterious gadgets—handles, buttons, and boxes of differing dimensions.

"I'm testing their learning ability and reactions," she explained. "This one on my right can turn on a picture show whenever he likes—colored slides, you know, one after another, on that little screen thing. When he gets tired of watching those, he can get himself a drink of water by pulling the string here. He can make this toy slide up and down, and open that inner door or shut it. He can turn on a light and turn it off again, and he can feed himself—within reason, that is. In other words, *he controls his own environment.*

"Now over here in the other cage the same things happen, but the monkey can't do them himself: they happen because the other monkey makes them happen. The environment of this second infant depends on the whim of the first one. He *can't* control his environment."

I carried away with me a vivid memory of the first monkey's face, which, though still infantile, was definitely smug. The expression of a leader.

The Washington Primate Research Center at Seattle is so closely intertwined with the state university that it seems—and is—even more a part of the academic building complex than that at Wisconsin. Dr. Theodore C. Ruch, the director, is widely known for his book *Diseases of Laboratory Primates* and for the *Bibliographica Primatologica,* which has developed into a nationwide and worldwide bibliographical service. Before becoming one of the founding professors of the medical school at Seattle in 1946, he spent some time at Yale in the ambience of the never-to-be-forgotten Robert Mearns Yerkes and John Farquhar Fulton: he says he learned there that, public prejudice to the contrary, monkeys *can* make good neighbors.

"They do not have to be consigned to a large tract of land in the country or banished to the roof or the basement," he said to an interviewer. Nevertheless it was against a good deal of opposition that he succeeded in putting through the plan on which the center is built, an "I" building that connects the basic medical science departments and medical and dental school complex on one side with a biochemistry-genetics building on the other. Within these long, long passageways, therefore, the biomedical sciences work together cheek by jowl—but an elaborate system of doors between them and the primate colony must be kept closed at all times, as if to repel some Wellsian invasion by well-drilled infrahuman troops. Outside, against the state of Washington's lush greenery, the effect is tidy and pleasing, and so is indoors—though grim dramas are sometimes played out in the monkey colonies housed in the laboratories. The walls have seen more than one breaching, for example, as in the case of the Japanese macaque troop that lived for a time on the top floor.

These monkeys, sturdy red-faced creatures stockier than rhesus and weighing as much as forty pounds when they are adult, live by choice in tight social groups under a dominant male, and the Seattle troop was no exception. It was stratified in the familiar manner, with a top boss, several promising underlings who kept an eye on their chances of promotion, females, bachelor males, and juveniles. One day a research worker came up to see how they were getting on, and found their exercise yard empty—or so he thought—for cleaning. Assuming that the monkeys were confined in their waiting yard, he wandered out onto the floor, only to be attacked suddenly by the Old Man, who had been lurking behind some architectural feature or other. Japanese macaques, as I have said, are sturdy, and they are also supplied with long, wicked teeth. The angry Old Man launched a flying leap at the intruder, who landed a punch on the monkey's chest, hitting him more or less by accident in the right spot. The Old Man fell stunned to the floor, and the human made his escape.

Before the Old Man had come to his senses, one of the eager young pretenders had taken his place. You might suppose that as soon as the erstwhile boss got his strength back the old order would be resumed, but no: from that time on, things stayed as they now were. The new king fought off the old one successfully, and after that one attempt the old one never tried again.

"He had lost face," my informant explained.

Just before my first visit to Seattle I had read an article by two people working at the Washington Primate Center, Dr. Gordon E. Jensen and

Dr. Ruth A. Bobbitt, in the May 1968 number of the magazine *Psychology Today*. The doctors explained in the introduction that they were starting at a point to which they had been led by the studies of Dr. Harlow and his wife Margaret, which demonstrated that infant monkeys deprived of maternal care, if they are brought up among normal peers, may still develop essentially normal social behavior. Doctors Jensen and Bobbitt studied another type of behavior, the mother-infant relationship. They did not use rhesus for this work, preferring the slightly larger pigtail macaque, *Macaca nemestrina,* the favorite monkey of the Washington Center. Needing a number of infants more or less the same age, they arranged that the prospective mothers should be mated at the same time. I noticed that they employed the same stratagem favored by the Wisconsin staff when the time came for birth. Monkeys usually give birth at about 2:00 A.M., a most inconvenient hour for Homo sapiens who want to be present on such occasions, so the pregnant monkeys were kept on a reversed lighting and feeding schedule, to make them think day was night and night was day. It worked. The mothers were persuaded to go into labor in what was actually early afternoon, and the attendants got their full night's sleep. At Seattle the researchers carried this system even further: not only did they reverse the day-and-night sequence and the feeding schedule, but during the true nighttime they kept a radio going in the laboratory, tuned to an all-night disk-jockey program. Throughout the night the pregnant pigtails were thus fooled into thinking that the world was awake and working, and cheerfully waited until an artificial quiet descended and lights were turned off to do their stuff. Jensen and Bobbitt, of course, were waiting, notebooks and pencils at the ready.

Soon they were able to record an interesting observation. The old idea of the monkey mother who just sits and clutches her child doesn't apply very long to *M. nemestrina.* Affectionate and protective at first, the mother soon starts pushing away the infant, insisting that he learn to get along by himself. Anyone who has watched how a cat, expecting a new litter, discourages her growing kittens from nursing will get the idea—but a pigtail mother is not pregnant when she chases her baby away: it is merely her way of life. One might say that she has other things on her mind: she can be described as a working-mother type. *Nemestrina* mothers, the researchers decided, are natural "leavers," whereas the infants, reluctant to accept this state of affairs, are "approachers." (Living in England, I have suffered quite a lot of disapproval from the natives, who say that my tendency to leave home is unnatural. I am seriously considering taking a troop of pigtail macaques back with me and setting them loose in the Hertfordshire woods. *That* should show the English what is natural and what isn't.)

But Dr. Jensen and Dr. Bobbitt had only begun their researches. "One of our goals," they wrote, "was to determine how isolation of naturally mothered infants would affect their development. We wanted to know if isolated, mother-raised infants without playmates, or with later contact with playmates, would develop normally. We arranged a study of mothered monkey infants according to the amount of contact with playmates of the same age, ranging from none at all, through visual and auditory contact only, to full and free association. If we could understand the importance in humans of contact with companions of the same age, we might be able to devise improved child-rearing guides for parents and programs for optional institutional and foster care of children. This study also showed the effects of deprived and enriched environments on monkey development."

They made up three different types of environment that would supply conditions of deprived, delayed-enriched, and enriched. In the deprived environment, mother and infant lived in bare quarters, cut off from contact with man or other monkey. In the delayed-enriched environment, mother and child were kept in a similar dwelling for four months but were then supplied with playmates and toys. In the enriched environment, the subjects had toys from the beginning, and their cage stood in a large open laboratory within sight and sound of people and other monkeys. After four months these rich baby monkeys were separated from their mothers and became playmates for the new-rich infants for two months. Then the whole lot of them were tested.

The writers report: "We found that the privation environment produced prolonged physical closeness between mother and infant, or what could be called a retardation of the mutual independence process. . . . In the privation environment, the infant and mother turned to each other for most of their reinforcements. This situation is analogous to the overprotective human mother and her child, or to families living in very isolated and severely deprived conditions." Nevertheless, the tendency of a mother to leave her infant, while the infant made most approaches, remained true in spite of the rearing environment. "These and other findings led us to conclude that there is a basic process to the development of mutual independence in laboratory monkeys. The process is continuous and begins in the first weeks of life rather than at a particular stage such as at weaning or later. A rich environment does not affect the nature of this process: it merely produces an earlier elaboration of the infant's behavior towards independence."

To compare the "social adequacy" of the three types of monkey, the infants were placed together in pairs. The researchers say, "It was evident that the richer the early environment, the greater the infant's social development." The privileged infants dominated the deprived infants even where the deprived monkey was male and the enriched monkey female. Privileged animals usually dominated monkeys from the delayed-rich environment, and they were never dominated by them even though, it will be remembered, the privileged infant was at a disadvantage, snatched as he was from his mother and his comfortable home and placed in a new home with a strange playmate and the playmate's mother, who was usually hostile. "That the rich environment infant did so well," say the writers, "emphasizes the advantage provided by his rich environment in the earliest months of life. Conversely, the retardation in social and motor development shown by deprived infants in the earliest weeks of life later manifested itself in the poor showing of deprived infants in social and competitive situations. . . . These studies . . . suggest the potential value of human environmental enrichment programs such as 'Head Start' and anti-poverty programs that provide educational and cultural opportunities for deprived school-children."

Little was said about the father's role in pigtail family life, the authors explain, because monkeys haven't the same monogamous family structure "popularly ascribed to humans." Males play an important part in the social organization and defense of the group, and no doubt they influence young monkeys' learning to behave as group members, but the subject of father-infant relationships, as well as leader-infant relationships, has still to be investigated.

One thing, however, can be said about the differences between the sexes in the monkey world. Human mothers often claim that boys behave differently from girls right from the start. Our writers have examined this question with pigtail infants and found that with them, at least, this is true. Male monkey babies do indeed develop behavior patterns that differ from those of female monkey babies. They achieve independence sooner, and they are generally more active. As for the mothers, they punish their male babies more than females, hold and carry them less, and pay less attention to them. "Failure of human mothers to 'let go' is often postulated as a cause of social problems in males," add the doctors darkly.

In conclusion, they say that their experiments have led them away from thinking in terms of the exclusive importance of Mother. "With all due respect to contemporary literature, television, films and funnies, that satirize and spoof mother, it appears that although mothers are important, their role has been oversold. Experiments support the idea that playmates and environmental factors are important too. In time, although mothers may get less credit for man's success, they may also receive less blame for his failure."

Dr. Douglas Bowden, a member of the permanent staff at the Washington Primate Center and assistant professor of psychiatry at the University of Washington School of Medicine, recently returned from three months in Russia where he worked for one month among the Russian primatologists at Sukhumi. He and his wife loved Sukhumi, which they describe as an ideal summer resort, much like northern California.

"Isn't it rather unusual for you to have stayed so long?" I asked.

He said, "I guess it is, but I was part of the Soviet Academy of Sciences Exchange Program. It was a fascinating time—I hated to come away. They've had a group of chimps under observation there at Pavlov's summer laboratory for years and years; only last year they broke it up and started getting some new ones in. All that was fascinating, too. One of those animals, I firmly believe, died of depression. Before that, she'd never been sick in her life: she'd been there ten or twelve years.

"There were five in the group—one male at least, as I remember, and three females, but I'm not sure of the sex of the fifth. It doesn't matter. . . . The Russians were about a year selling them off or swapping them one by one. It was noticeable that the chimps' behavior changed with each departure. The one I'm talking about, though, wasn't sold—they decided to keep her because she was the easiest to work with. She'd been dominant in the group for years and was the male's closest companion. Well, she was okay until the male was gotten rid of, and then she became quite a different creature: depressed, unresponsive, and with no appetite —clearly she was really physically ill.

"They did all kinds of tests but couldn't find a trace of infection. It went on like that for several weeks until at last she developed a gastro-intestinal disease, and died of it. In the autopsy all they could find as an immediate cause of death was an infarction of a mesenteric vessel—a disruption of the blood supply to a portion of the intestine. *I* think it was a genuine case of heartbreak. Of course, one hears many stories of animals dying mysteriously in captivity, but this was different. She'd lived in captivity a long time."

Dr. Bowden has had his turn at studying depression in laboratory monkeys, and is interested to find that different species show very different emotional reactions to such supposedly commonplace situations as motherhood: "Langurs, for one, haven't the same attitude toward their children as macaques. In a group of langurs, children are treated like common property: everybody looks after all of them. Presumably you'd find it pretty hard to induce a depression in a langur. As you have no doubt heard, we're investigating the possibilities of other primates than rhesus for lab research, and one of the interesting lines I followed a few years ago at the Max-Planck-Institute in Munich was connected with *Saimiri sciureus,* the squirrel monkey. There was an idea that they wouldn't be much use in research because they don't breed well in captivity, and we tried to find out the reason. We noticed that breech delivery is a lot more frequent in monkeys than in humans. Then I've done some work on the hormone, or proteohormone, prolactin—it's found in the anterior pituitary. It's what stimulates lactation. We've found that the level of prolactin in rats is directly related to the rat's maternal instinct, so called. Even a male rat injected with prolactin will feel maternal—at least that's the way he acts.

"By the way, talking of Sukhumi again, I just remembered a story they told me about a special experiment they tried out on one of their male chimps some time ago, before I was there. It was a kind of intelligence or adaptation test. First they trained him by lighting a little fire, taking a bucket of water, and showing him how to put the fire out by pouring the water on it. Then they gave him the bucket of water, lit the fire again, and let him do it all himself. He learned the action quickly. So far so good. For the next phase, they put him with an empty bucket on a raft on an outdoor pond near the laboratory, lit a fire on the raft, and shoved the whole contraption away from the bank, the idea being, of course, to see if he was smart enough to dip up water in the bucket for himself and put the fire out. What do you think happened?" Dr. Bowden laughed. "He put the fire out all right—he urinated on it."

Dr. Dwight Sutton was a visiting scientist, an associate professor in psychology at Arizona State University.

"I'm spending my sabbatical here," he said. "I'm working with another man on the function of the auditory area of the cortex. Monkeys can hear noises all right, but often they don't seem to care about them—they pay no attention unless the hearing is reinforced. It's not deafness, but there's no doubt one can't get through on the proper channels. Sometimes a noise *will* register momentarily, but if it's repeated, it's ignored. We ask ourselves, therefore, how the cortex works, exactly—How does it fade out sounds? Because that's what the monkey cortex does; it edits, or damps, sounds. We haven't got a precise, valid picture as yet, but I think we're on the track.

"I'm also doing a study of muscular control, more explicitly the mechanism of tremor, for which I'm using stump-tailed macaques which I brought along, because they're comparatively tractable. You can train a monkey to hold out his hand, you know, and keep himself quiet in that position. If he's reinforced with applesauce, he gets very skillful at doing this, very quickly. A stumptail is as good as a man at this trick, though he's slower at the beginning, usually, to learn to do it. But if he's poisoned with a drug, for instance, amphetamine, he gets shaky—as do we. Another drug that makes him shaky is tremorine; in fact, it produces a larger tremor like that of Parkinsonism. Drugs can be used to elucidate the nature of the phenomenon, to evaluate it. Some suppress, others reinforce, it. Then there is the question: If the motor part of the cortex is removed, how much would the operation impair muscular control? We haven't analyzed that problem as yet—we're still on the pharmacological analysis. Obviously there are lesions that can mimic the symptoms of Parkinson's, and when we know more, it might be possible to make a very early diagnosis of muscle impairment, before it shows."

For a time, I watched a complicated apparatus in which a rhesus monkey, gobbling applesauce, pressed buttons to get more and more of the sauce, while all the time an electrode in its cerebral cortex moved very slowly, searching as it were for a firing cell. The activity of the cell showed on an oscilloscope along with a record of the monkey's eye movements. Above the oscilloscope was a large screen like that of a TV set, which showed an enlarged picture of the monkey's hands busy with the buttons. He would continue to work, the investigators told me, until he had had his fill of applesauce.

With my guide I paused as we went toward the elevator to look at a very large hamadryas baboon, the largest I had ever seen. "Beautiful, isn't he?" said my companion. "He weighs about ninety-five pounds. He's on a high cholesterol diet as part of our studies of cholesterol transport and its involvement in certain types of cardiovascular disease."

Down in the lower regions, Dr. Daris R. Swindler presided over an office with crowded shelves and many diagrams. He is a research affiliate at the center and also a professor of anthropology at the university. He was working, he said, on an atlas of the comparative anatomy of man, baboons, and chimpanzees.

"I'm also sudying the dentition of all living primates from the tree shrew up," he said. "New World primates have three premolars, therefore more teeth than Old World primates. Did you know that? Old World primates, of course, have three molars but only two premolars, and so have we. Now, marmosets have lost their third molar, at least it's very rarely seen in marmosets, but like other New World primates they still have their three premolars. I'm computerizing all the data I have on dentition, so that I can define the parameters of variability in extant species. This information will help paleontologists to understand better the dental variability that existed in fossil specimens.

I had to get up early in the morning to catch a plane to Spokane, where an outlying part of the Washington Primate Center is to be found at Medical Lake. I wondered at first why the center should have gone three hundred miles away for the necessary space on which to establish a field center: it's not as if Seattle were like New York, I reflected. On the contrary, one of the city's chief beauties is that there is plenty of space nearby. On the plane, however, Dr. Orville A. Smith, Jr., the assistant director of the primate center, explained it all.

"We were awfully lucky to get this place. It was built as a maximum-security prison for the criminally insane, as part of the Eastern Washington State Hospital, and I guess it was to soothe public sensibilities that they put it way off where it wouldn't scare people. The state must have spent millions on it. Then, soon after it was ready to open, tranquilizers were discovered. Bang!" He struck his hands together. "All of a sudden there were few patients requiring such prisonlike restraint. It was dramatic, I can tell you. It was wonderful, a real blessing for humanity—but there the hospital was with a brand-new prison on its hands that was no use to them. It would have cost too much even to pull it down. In the end it was handed over to us for a field station for breeding colonies of monkeys. Other colleges use it with us: we have arrangements with Eastern Washington State and a couple more. It's been a godsend."

A car that met us at the airport took us straight to Medical Lake outside town—a much prettier place than the name implies, with trees and quiet peace prevailing. We circumvented the mother hospital and drove a fair way inland until we arrived at the former prison. It did look like one, no doubt about it, with its concrete and the bare ground around it.

"We're going to do something about this," Dr. Smith said, "just as soon as we've decided whether or not to keep some animals outdoors."

A man in overalls met us and led us through high corridors in a brown sort of gloom—not the man's gloom, for he was cheerful; it had something to do with the building itself, and the lights that didn't seem strong enough. Here and there walls were punctuated by metal-sheathed doors that had to be unlocked for us, not so much because there was need as that it was the only way the doors worked. After all, criminally insane prisoners. . . .

"They had a system that locked everything at once," Dr. Smith told me. "The guard would sit there, in that circle, and keep an eye on things. If anyone had escaped, the doors would have been shut automatically. Depressing, isn't it?"

Just then we walked past a row of little cubicles that stretched the whole length of one passage and opened out into it. Each had a high, small window, barred.

"No more depressing than a girls' boarding school in England, except for the barred windows," I said. "In fact, it's a good deal like a school, but where are the pupils?"

"That's just it—we've got so much room they're hardly noticeable unless you go and track them down." Dr. Smith lowered his voice. "This man, Jim, who's taking us around; he's absolutely wonderful with baboons. He loves them: they're his life. And to think that we'd never have known it if we hadn't decided to breed baboons as well as pigtails! It was what you might call a hidden talent: he didn't know about it himself until the time came."

We were approaching familiar noises, and a smell I have grown to recognize, of a particularly pungent antiseptic. Turning a corner of the corridor we came to another row of cubicles that had been slightly altered; they had barred doors and—between each two—a little runway at the bottom of the wall, from one room to the other. There were floor drains, too. A group of pig-tailed macaques disported themselves in the first suite we stopped at—at least a dozen of them, chasing each other, slipping through the hole into the other room and back again, and generally being very busy.

"This is a breeding group," said Dr. Smith proudly. "They seem to have taken to the life here, and they're doing fine. We've got another lot down here at the end, and over there are the baboons. Down on the next floor are crab-eating macaques—and, of course, rhesus."

"Of course."

We peered at room after room. "It's going to be a practical proposition. I'm sure it is," said the doctor. "We've been working on the cost of breeding and it cheered us up a lot. Just think; if you get a female pigtail from Asia and count everything in the way of expense—initial cost, transport, quarantine, treatment—a baby from that animal will cost about six hundred dollars. Furthermore, you'd know nothing about the infant's pedigree. Now, when we've got this place running to the limit of capability—I figure we could take care of three thousand animals instead of the few hundred we have now—well, barring accidents and acts of God, I figure that a pigtail infant from our breeding group, free of parasites and with known background, shouldn't cost more than two hundred and fifty. And we won't be depleting the wild."

In a nursery room a large number of prison-bred pigtail youngsters were playing on the floor, on plank perches, and up and down some shelves that had been put along the walls. All were more or less of an age, with one exception—a large, brooding female who sat in the middle of the room with her arms full of infants. Other babies hung around her neck, crowded on her back, and plucked at her arms. She looked at us amiably above the three or four furry heads pillowed on her breast.

"What on earth is that?" I demanded of one of the caretakers.

"Well," he said, "she just has a talent for it, that's all. Never had any of her own, but she looks after all the young ones and seems to like it, so we let her stay there. As a matter of fact, she's a great help."

6

Binoculars and Microscopes

Beaverton, site of the Oregon Regional Primate Research Center, is so near Portland that in any less fortunate state it might long since have been engulfed by unlovely urban outskirts of that city. It is hard to spoil Oregon's beauty, and certainly the center adds to it rather than detracts. The first of the Regional Primate Research Centers—dedicated in May 1962, two years after the initial grant was awarded—it lies in a country setting, artfully landscaped, with plants and flowering shrubs lining the pathways that lead from one building to another—as unlikely a place as one can imagine to harbor Japanese macaques and African bush babies.

Dr. William Montagna, the director, formerly professor of biology at Brown University, picked me up at the hotel to drive me out to Beaverton, and on the way talked at some length about cutaneous biology and baldness, related subjects that interest him professionally as well, no doubt, as personally—for Dr. Montagna, tall and lean and full of vitality, is bald.

"There's one species of macaque, the stumptail, that becomes bald at adolescence. That's a fact," he informed me, deadpan. "Remind me to show you some at the center. Adolescent and adult animals show a receding hairline, and baldness progresses from there. You know studies of baldness interest a lot of people—cosmetics people, for instance. We've had grants to study it."

He talked of other things, too, such as the center's unique collection of prosimians, which—as the name suggests—are animals a little lower than monkeys on the family tree: The term embraces lemurs, galagos, and bush-baby types. Beaverton's collection represents all but the rarest of the true lemurs, which are found only in Madagascar—or, as that country is now called, Malagasy—and a large number of lemuroids from other countries. Dr. Montagna also talked of the center's relations with Beaverton's public (good), and various projects on which the staff and visiting scientists are engaged. But when, that evening at the hotel, I read an issue of *Primate News,* the center's house magazine, I was pleased to see that in a column called "From the Director's Desk," Dr. Montagna had reverted to his favorite subject, hair:

> I have for some time viewed with indulgent amusement the increasing number of men at the Center smitten by the revival of ancient hair styles [he wrote]. New beards appear daily, long sideburns and muttonchops are becoming paramount, and flowing hair will soon come. When I remarked about this to a colleague who sports fuzzy sideburns, he replied: "Women like it." Yet, there must be a more profound reason.
>
> Since only men can grow a beard and mustache, doing so is an assertion of masculinity. But here is a conundrum. Scalp hair, lush through adolescence, in most individuals becomes progressively anemic as the hair elsewhere on their bodies attains greater robustness. The same genetic factors that control the attainment of a hairy chin and chest also promote the enfeeblement of scalp hair as a man becomes progressively mature. For maximal expression of secondary male characteristics, then, I propose a new fashion: wear long beard and mustache, design clothing without a front, exposing one's hairy chest to full advantage, and shave off all scalp hair. That will surely give the stamp of unquestionable vitality. . . .

Dr. Montagna does not, however, spend all his time thinking and writing about hair. His latest publication, written in collaboration with Richard J. Harrison, professor of anatomy at Cambridge University, England, is a book entitled *Man,* the nature of which some of the chapter headings give an idea: "His Place in Nature," "Man's Curiosity About His Anatomy," and "What Kind of Creature?" As to this last, the authors quote another doctor, George W. Corner, who said, "After all, if he is an ape he is the only ape that is debating what kind of ape he is."

Two features of the center that attract the layman's eye are both, as it happens, Japanese in origin, though that is their sole similarity. One is the work of Joel Ito, who as a boy in Los Angeles studied zoology, later majored in art at the University of California, and then went on to take an M.S. in medical illustration. Few people in this country go in for medical illustration, and Ito had a wide choice of jobs. He chose the Oregon Primate Center. Now he leads a double life, professionally; he makes strictly medical drawings but also produces portraits of primates in watercolor, carbon pencil, crayon, or etching—whatever medium he feels suits the subject. Both classes of picture, medical and portrait, by Joel Ito hang on most of the center's walls, and he has illustrated a 1970 calendar with a different primate for every month, which is one of my favorite souvenirs.

The other feature I mentioned is the Japanese macaque colony living outdoors in a large corral on the center estate. This group was captured entire, just as it was living in the wild, complete with leaders, females, bachelors, children, and all. It was brought wholesale to Oregon, where the members continue doing whatever Japanese macaques do under the eyes—sometimes aided by binoculars—of spying behaviorists far above them in one or two observation rooms. They are creatures of habit, and have marked out their corral in definite areas: one where the young ones play, another for feeding, another for sunning, and so on. Dr. Bruce Alexander was watching them when I climbed up to an observation post, making notes on a chart as he did so. He pointed out the markings on the animals that help people to distinguish individuals—dominant males, for instance.

"They don't always behave quite as we expect them to," he admitted, "and it may be that they're altering their habits here. Females tend to squabble, and that's natural enough, but in the wild when this happens, the male breaks it up, whereas here our males don't do that very much. Japanese macaque males help to mother the babies, you know; possibly this isn't true of the rhesus, I'm not sure. The main structure of the group, though, resembles that of the rhesus and the baboon: it's a family structure. You'll see siblings defending each other and defending their mothers—and vice versa, of course. One family here has two siblings that aren't related—one of the infants was adopted—but they behave exactly as if they were really in the same family. Though most youngsters in the same age group will defend each other if they're threatened, the defense isn't as strong with nonsiblings as with siblings. I should say that the strongest defense of all is that put up by a mother for her offspring."

He showed me several cages standing near the corral, which he said were being used for a certain type of experiment: they were putting groups of the macaques into small cages to see what crowded conditions did to their behavior. For one thing, he said, the troop structure held up very well in the crowd, even more than they had thought it might. It remained consistent. But individual males show increased aggression, sometimes even mobbing each other. Reducing the number of females to males, on the other hand, didn't seem to make the males any more short-tempered.

Whether or not they behave the way they ought to, Oregon air seems to suit the Japanese monkeys, for they breed well enough to threaten a possible problem one of these days of overpopulation. In the wild when a troop reaches the number of one hundred—more or less—it splits as if by common consent, a subleader taking the split-off group to another part of the forest. The Oregon colony at the time of my visit numbered eighty-four.

"When they decide to divide up, where are the extra ones going?" I asked.

Dr. Alexander said, "We haven't figured that out. After all, they aren't yet at the fissionable stage."

"This has been a good year for Cutaneous Biology," said the center progress report, so I investigated that department, where Dr. Mary Bell showed me pictures she had taken with the electron microscope of sebaceous glands in various primate species. These glands are often attached to the hair follicles—in us as in other mammals—and they supply the oily substance sebum that keeps hair from drying and falling out, or rather, from breaking off. In the first pictures Dr. Bell showed me the glands—enormously magnified, of course—that looked like lakes or peninsulas; they were irregular areas surrounded by smaller details on the map of some mysterious country. But that was only an introduction. Photos of much more highly magnified glands exhibited an amazingly regular pattern at certain places inside the boundary lines. These patterns, or grids, are structured elements called lysosomes.

"Here's one that looks like tweed," said Dr. Bell, truthfully, "and here's another that looks like crocodile skin. The most surprising thing we've found out is that all these patterns are different. You get a variation for every species. Once you recognize one you can always be sure that it came from a certain animal."

In the Biochemistry Department I met Dr. Clarissa Beatty and her associate Dr. Rose Mary Bocek, who were working on four major projects. They talked first about the metabolism of fetal muscle, which they were studying as part of a larger piece of work on muscular dystrophy. Dr. Beatty said they were focusing on the intermediary metabolism of carbohydrates and lipids—fatty acids—in fetal and newborn muscle, which nobody had investigated before.

"We're using the fetus of rhesus monkeys, and we've found out a lot," said Dr. Bocek.

"We knew that a neonate uses lipids in the metabolism of its muscle," said Dr. Beatty. "Now we've gone further and found that the fetus, too, utilizes fatty acid. In fact, it seems to use more in proportion to its weight than the adult does. The lipid is transmitted by the placenta to the fetus, just the way glucose is. We think it might be used for energy, though we haven't proved it. The assumption used to be that muscle utilized only carbohydrates, but we know now that it uses some protein as well. We've also been looking at the effect of insulin on fetal muscle. As is well known, insulin has very much of an effect on carbohydrate metabolism in adults, but nobody had tried it out on fetal muscle until now. The results we've been getting indicate that the fetal metabolism is responsive in the same way, and of course this is well worth knowing if it's true. Our studies of muscle, as we said, lead to work and more work, since we want to know the causes of weakness in voluntary muscles that is called muscular dystrophy."

Doctors Beatty and Bocek have been trying to induce this disease in monkeys. "Nobody's succeeded so far," said Dr. Beatty. "As you probably know, it's sex-linked. Women are carriers, though often it's not easy to see in them: you need blood tests to identify a carrier. A type of dystrophy has been developed in certain animals—in chickens, mice, pigs, even cats—but it's not near enough the human kind to be much use. If only we could start it in primates, that would be so useful! Their blood types are similar to ours; they have heart attacks like ours, and strokes; and they get viruses like ours—many of our maladies. And their muscle structure is very much like ours."

She added that the Muscular Dystrophy Association was helping them on this research financially.

I noticed that both doctors used the terms "red muscle" and "white muscle," which meant nothing to me, and I asked what they meant. Dr. Beatty explained. "The voluntary muscles, which are the ones dystrophy destroys, are made up mostly of red fibers. White muscles are those that act quickly and tire quickly: they're made of mixed red and white fibers."

"Think of a chicken," said Dr. Bocek. "The drumstick is red muscle, and the breast is white."

"Simply comparing the metabolism of these two kinds of muscle leads one to a lot of new knowledge," said Dr. Beatty, and both ladies smiled contentedly.

After this conversation it seemed a natural transition to listen to Dr. Oscar Portman, of the Department of Primate Nutrition, on the subject of the pathogenesis of atherosclerosis. This disease is a phase of arteriosclerosis characterized by the deposition of a fatty, waxy substance on the inside walls of the arteries, so that the blood channels become progressively narrower—choked up—and the vessels themselves harden and lose their elasticity.

"It's a disease that starts very early in life," said Dr. Portman. "People tend to think of it as an old-age symptom, not realizing that it's really a pediatric disease, begun by the time the subject is in his teens. It doesn't show in youth, but it's there—it's there: it merely grows more severe as time goes on, and usually doesn't show up until the third or fourth decade of human life. The idea that atherosclerosis might be connected with diet began with epidemiological studies during which it became obvious that certain races are prone to it and others aren't. I won't go into detail, but we can safely say that the Japanese peasantry, for instance, who eat sparingly of meat and subsist largely on rice and fish and pickle, suffer very little from atherosclerosis, and the same is true of the people of India, who also live on a spare diet. Atherosclerosis is a disease of prosperity—underfed people don't usually suffer from it, though all races do exhibit at least the occasional case. Well, all this started us thinking that we should study certain aspects of diet and its effect in this respect on young nonhuman primates. We asked ourselves, What are the mechanisms that change the lipid composition of arteries with the approach of age?"

The word lipid means fat or fatty. Lipids of various kinds, with proteins and carbohydrates, are the principal components of living cells, and in the arteries of an ageing person afflicted with atherosclerosis they are present in a disproportionately large amount. One of the problems of studying the disease in a laboratory animal is that the research worker can't just sit down and wait for his subject to grow old, so Dr. Portman and his associates speeded up the process with diet. They were using two kinds of primate: rhesus and squirrel monkey. Twenty-four rhesus were fed from birth to the age of a year on various formula diets that are used for human babies. It was found that those animals fed on formulas containing butterfat showed a much higher concentration of lipids in their blood than the ones fed on corn-oil formulas.

"But none of them has yet developed atherosclerotic lesions," said Dr. Portman. "We'll have to wait. On the other hand, when we feed young squirrel monkeys an atherogenic diet—that is, a diet rich in eggs and milk—we find that this results in chemical changes in the arterial wall which we know are forerunners of the disease. These chemicals speed up the ageing process, and lipids, especially the phospholipids, tell a nice story as they undergo changes. To observe these changes we have several methods: for one thing we can strip the lining of the artery for microanalysis, but it isn't easy because a squirrel monkey's so small his whole artery doesn't give much material.

"Everything considered, we think we can begin now to test methods of reversing the process, curing the developing atherosclerosis, but how?

The obvious answer, using a corrected diet, has been tried, but as far as rhesus are concerned, at any rate, it isn't promising."

"If we had to depend on humans for trials of drugs to alleviate coronary atherosclerosis, we'd need to dose ten thousand patients for five years with one drug," said Dr. M. R. Malinow, chairman of the Department of Cardiovascular Surgery and associate professor of medicine at the university medical school. "Added to which we'd need controls, of course —at least five thousand of them. Quite apart from the dubious ethics of the exercise, we would spend thirty or forty thousand dollars a year during that time, which is why it is vitally necessary to use nonhuman animals in research; it all goes much faster for far less expenditure. Now at last we have found a species that shows lesions—the crab-eating macaque. Lesions develop after only three months on a special diet containing cholesterol, butter, and sugar. We have the species; we have the method; now we are trying out the drug. We are trying out *a* drug, that is to say —so far there's no hard data on it as used on humans. If it has no effect on monkeys, doctors won't use it on humans, but if it alleviates the disease in the monkey, that will be different.

"Muscular energy increases the degradation of cholesterol. This is not news. Exercise, taken in time, is good for you. We have seen the process in rats, who if they run, burn up more cholesterol, but we have not been sure what organ is responsible for this action. So I removed the liver from a rat—the whole liver, you know, and that is not as simple an operation as it sounds: I had to go to the Mayo Clinic for the right equipment to do it. Afterward the rat was still able to increase the oxidation of cholesterol by exercise. Plainly, then, the liver was not the organ we sought. What other organ? I made a guess—an educated guess, I admit —and removed the adrenals. And that was it: the rat couldn't burn up his cholesterol any more.

"Now, how does this work? How do the adrenals know that the muscles are contracting? There are two possible methods—by electrical impulses through the nervous system, or by hormones. Almost certainly it's not the nervous system, so it must be the hormones. But no known hormone is produced by the muscles. We are still working on this problem."

Hay-fever and asthma sufferers may have already heard of "the timothy hay treatment." It was evolved at the Oregon Primate Center by Dr. Arthur Malley, of the center's Department of Immunology, and Dr. Frank Perlman, chief of allergy studies at the University of Oregon Medical School. They used rhesus monkeys for their research.

"There are some basic principles common to all allergies," said Dr.

Malley. "We study why an allergy develops, what sets it off, and how to prevent the reaction, which might be tissue damage or even death—for death does occur in some cases. Exposure to environment often sets off the allergy, and treatment has been based on the fact that the patient must take the materials that cause the reaction in order to purify his system. I've been working on this problem for thirteen years. Now we've isolated a very small fragment of a molecule. This particular molecule, as it happens, is for grass pollens, but the principle that underlies the treatment is also applicable to many other substances. The fragment can combine with the irritating body and deactivate it.

"Hitherto, treatment for hay-fever patients has consisted of injecting a pollen extract which produces blocking antibodies in the patient's system and protects him against subsequent exposure to the irritant. The trouble with this method is that other antibodies called reagins, already present in the patient, react with the protein allergen in the pollen: this releases histamine and produces the hay-fever symptoms that we wish to avoid. In fact, an overdose of allergen may prove fatal. Thus, crude pollen extract must be administered in very small doses over a long period of time: it might be three years or more before the patient gets any significant benefit from the treatment. We've been using our fraction of a molecule for only three and a half years, and it seems to work.

"People lose their antibody titers very soon [a titer is a measure of the strength of a concentration]: the titer drops early in the treatment and stays low for the rest of the time, which suggests that we are purging the patient's body of the antibodies. In our trials, for six months, from August through January, none of the patients received treatment after the first series, but the reagin antibody levels remained at the same low points we had reached during the program. Patients who had become desensitized to timothy pollen didn't slip back into their allergic sensitivity. After six months they evidently began making the antibody again, so we resumed the treatment. Well, after three years some of them don't begin again at all. A complete cure, then, is not impossible."

What might be even more important in this work, continued Dr. Malley, is that the principle also applies to the chief difficulty encountered in organ transplants—the rejection factor we have all heard of. This rejection is an immunological barrier, or in other words an antibody reaction, in which the antibodies automatically rush to defend the body against the foreign influence of the newly introduced transplant. To slow down these rejection phenomena, if not to prevent them entirely, surgeons have to use immuno-suppressive agents—various chemicals such as immurand or cortisone, and these, when used for a long time at a stretch, have undesirable side effects. The white blood cell count falls, and the patient's resistance to infection is lowered to danger point. In Dr. Malley's laboratory they have used rhesus white blood cells to isolate the specific antihuman lymphocyte antibodies from the other serum proteins, and have then administered this purified antibody to human patients suffering from autoimmune, or rejection, diseases. Indications are that this treatment eliminates the secondary complications one gets with nonpurified material.

"Rhesus monkeys have been invaluable in this research," he said, "because their tissue reacts so much like ours, and they produce the same antibodies that cause allergies in humans. We couldn't do the work with nonprimate animals."

"How does hay fever affect a rhesus?" I asked.

"Much as it affects people," said Dr. Malley. "The monkey sneezes a lot and looks miserable."

Always suggestible and often allergic, I too sneezed, and wiped my eyes.

Dr. Robert W. Goy and Dr. Charles H. Phoenix of the Department of Reproductive Physiology both started their careers as psychologists, the former at the University of Chicago and the latter at Boston University. After various posts they both went to the University of Kansas, where they met. In 1963 they came to the Oregon Primate Center. Now they work on hormonal studies and behavioral processes, which they consider a natural development of their interests.

"For two years we've been known as anatomists rather than psychologists," said Dr. Goy cheerfully, "but the work we've done dates back to our Kansas days, when the first definitive report was made on the behavior of animals that had been treated as fetuses with testosterone. In Kansas we did this study with guinea pigs and rats. The conclusions were provocative, but they weren't as well accepted by our colleagues as we would have liked. Still, we understood the reaction: it was reasonable. Guinea pigs and rats seemed remote from our own physiology.

"About that time three people at Johns Hopkins were gathering information on the psychology of human hermaphrodites. The sex orientation of the subjects seemed to depend a lot on the way they'd been reared—that is, as male or female. And there was work being done on a certain type of habitual aborter; if such a woman was given synthetic progesterone it sometimes helped; presumably, then, her hormones were not normally balanced. Phoenix and I felt we must do more of our Kansas-type research, this time on species nearer to us—on primates. We'd watched Harlow's remarkable work on the normal play and the social and sex behavior of the rhesus, and we felt that this might be the break we'd been looking for. And Gertrude van Wagenen had been doing some pretty exceptional things—she'd given male hormones to young female monkeys, a treatment that had a profound effect on their morphology.

"We took pregnant rhesus and injected the fetuses with a masculinizing hormone. Out of this we got eight infant females that show male features of various sorts. How does the hormone do it? We haven't the definite answer, but in theory it works on the brain, somehow making some experiences rewarding and others not. Possibly there are centers in the brain that govern these reactions. We don't see them as pinpoint places but as something broader. We do know, of course, in a large way, that there are 'reward centers' and that the learning process depends on the reward system.

"These hermaphrodite females of ours behave in many ways like males. They use the threat gesture far oftener than ordinary females, though not as often as males. They use mounting behavior—normal females mount sometimes, but infrequently, and they don't adopt the truly male posture, but our females do. Their behavior in this respect is indistinguishable from that of males.

"We learned in Kansas that a masculinized female rat never ovulates, which makes one ask oneself what difference there is between the male and the female brain that regulates this function. Humans when given androgen do ovulate. So do our monkeys, but there is one difference: normal female monkeys begin their menses at from twenty to twenty-six months of age, whereas these pseudo-hermaphrodites don't begin until they are thirty-two months old. Evidently profound masculinization of the brain-pituitary-gonad axis does affect the clock.

"One would expect that a different kind of hormone would have different effects. We haven't used as large a dosage on the monkeys as we did on the guinea pigs in any case, nor have we given them testosterone as yet, but soon we're going to, after we've ovariectomized them. So far the penis of each animal has remained infantile, something like that of a castrated male; we want to find out how much further they will develop masculine traits without ovaries and with intensive treatment of male hormones.

"Incidentally, their body development doesn't go along with their masculine nature, and with their characteristically small female stature combined with masculine behavior, they're going to be in trouble. Each animal is in a social group: they have girl friends and all that. If we could have formed a group with some normal males and females and a few pseudohermaphrodites, that would have been ideal, but we didn't have enough of them for that. As it is each animal is assigned to a certain group at birth—it remains with its mother the first three months, then joins its social group for the next hundred days, where it is intensively studied. After that, it spends fifty days a year with the group, and the rest of the time is spent in various studies—two pseudohermaphrodites together, and so on. They're hardly ever alone. The monkey is a social animal, so each of our pseudos is allowed to play with a number of other monkeys at a time. At these playtimes she isn't closely watched, but anything important is recorded—big fights for instance, or changes of partners. Thus we have records on all the subjects, not only the pseudohermaphrodites but the others—a large number of normal monkeys, all completely recorded. This kind of thing is invaluable.

"The oldest of the pseudohermaphrodites is now eight. We have a pretty specific background on all of them, and this is of the utmost importance for psychiatric and psychological problems. With all this knowledge we can study many details of psychosexual behavior that were closed to us before. We have already noted a very gradual but decided development from pansexual [both homo- and heterosexual] to heterosexual behavior in normal males. One question we'd like very much to get the answer to: it deals with autoerotic behavior. What does a rhesus mother do with her male infant to discourage autoeroticism? Certainly she does something, but we haven't been able to find out what it is.

"In all we have about 150 animals—that is, including normal and all. Now and then we have to add a few new groups to keep it up to strength. We can't afford to sacrifice the well-recorded animals." (Here Dr. Goy was using the word "sacrifice" in its esoteric, laboratory connotation, meaning "to kill in order to examine." I find it an expression fraught with guilt, but no doubt scientists have discovered that such euphemisms are necessary.)

"To avoid sacrificing them," continued Dr. Goy, "we tend to shun electrode experiments on the brains of our pseudohermaphrodites—though it would be nice to find out what's going on there. After all, each of those animals is practically irreplaceable. Eight years' work, observation, recording—no, one couldn't replace any one of them."

I remembered these words after I had left Beaverton and was reading the copy of a letter Dr. Montagna had written to a friend at the NIH headquarters in Bethesda. He had just heard that the Division of Research Resources, as part of the government's much-heralded economy drive, proposed cutting the center budget for 1971 to a sum of $8.1 million to be spread among all seven centers. Dr. Montagna was appalled and dismayed.

"Directors and Administrators of Primate Centers cannot be dispassionate about these matters. . . . Each Center is threatened, and each must strive to survive against adversity and competition," he said. ". . . We have concentrated on attaining the highest standards of research both technically and creatively and on organizing an administrative team that functions harmoniously and effectively. Since nearly all of our programs are interdisciplinary, scientists of different disciplines work together. . . . Reductions in financial support would force us to alter standards, reduce productivity, jeopardize irreplaceable animals, damage the morale of our staff and, therefore, probably lose it—in short, reduce us to ineffectiveness." Summing up, he said, "It was agreed even in 1969 that $10.5 million was insufficient to support seven Centers; a reduction to $8.1 million in 1971 appears catastrophic."

Dr. Montagna therefore suggested, on the part of himself and his associates, a choice of several measures: to shut down four of the seven; to let the NIH take over the operating costs of certain centers—though this, he admitted, would be difficult; or, finally, to return to the original concept of one Primate Research Center instead of seven. Dr. Montagna made no secret of which one he felt would be the best choice for this role. "The Oregon Center has attained maturity and world-wide scientific stature," he said firmly.

The New England Regional Primate Research Center—I would be the last to blame those who refer to it as NERPRC—is the youngest institution of the seven, having been dedicated in 1966. It is also, I dare say, the hardest to find. Its address is Southborough, Massachusetts, but it is miles out in the country, disguised by woods and an occasional bristling sign that forbids the public to hunt, or reminds one that one is on government land. All this, the staff assured me, had nothing to do with the center except that it was built where it was for that very reason—it was on land already owned by Harvard University. No monkey is in danger of being shot. In fact, the center specializes in the smaller, more delicate primates, those that couldn't live out of doors in the cold north, though the center hall features a handsome half-circular glass room in which hardy young rhesus and other macaques romp, leaping about from various pieces of equipment to the glass circumference, where sometimes they sit and stare for a while.

"They're quite an attraction, aren't they?" said Dr. Derek Denny-Brown, who during the director's absence was doing the honors. "Yet I'm sure the architect didn't intend it in that light: he wasn't thinking of the place as a zoo. These monkeys don't live here permanently. Most of them participate in one experiment or another, and this place serves as a sort of recreation room between times."

Leading the way to the reception room, he continued, "The center is hosted by Harvard, though five other nearby universities and colleges are interested as well. We have a lot of back-and-forth with the Harvard Medical School, but there's also Tufts, Boston University, and so on—in short, New England. I myself was formerly at the Harvard Medical School, and before that, collaborated with Sherrington on neurophysiological studies of nervous mechanisms and disorders of primates—my chief interest is the nervous system. Our director, Dr. Bernard F. Trum, is another Harvard man, but he's been in other places, too: he was chief of the Veterinary Section in Europe during the Second World War and later with the Atomic Energy Commission."

Coffee was brought. "As far as numbers are concerned we aren't one of the larger centers," continued Dr. Denny-Brown. "We have only about seven hundred animals. We do a lot of work with the squirrel monkey, the saimiri: about one sixth our animals are saimiri. In spite of being so far north, by the way, we have a good breeding record for both rhesus and crab-eating macaques, but the New World monkeys don't do so well: they never do. One hopes to find out why. I doubt if it's the climate—outside climate doesn't make all that much difference nowadays, though one would expect it to have some bearing on one of the studies, which is to determine whether rhesus are seasonal in the strict sense of the word. Have they a seasonal sexual cycle? Sometimes one is led to think so, when they produce their infants more or less all at the same time of year; there is another possibility, however, that they can keep sperm in suspension until they're ready to let the infant start growing. At any rate it seems to have little to do with the true climatic season.

"We have a lot of visiting scientists here—as a matter of fact, they substantially out-number our core staff. Of course, there's also the permanent veterinarian and service department—probably the most important part of the center. Dr. T. C. Jones is our geneticist and Dr. R. Hunt, our pathologist. Dr. Arthur T. Hertig's chairman of the Division of Pathobiology and has recently been working on the ovary of the squirrel monkey. What might interest you most about our work are several special subjects: study of chromosomes, viruses in some New World monkeys—we've recently found a few interesting things there—and, in my own line (which you can hardly expect me not to talk about), a genuine case of cerebral palsy in a rhesus infant born here. And I should mention a study that is being made on the development of vision in young monkeys. Perhaps you know what has been done with vision in cats? If a newborn kitten's eyes are sutured and kept closed for ten weeks and then opened, the kitten never sees well afterwards. It has perfectly good equipment, good apparatus, but the sight degenerates, nevertheless. We find the same reactions in monkey eyes, here and at the Harvard Medical School. Kittens, of course, are born with closed eyes, whereas primates are not, which makes a difference in the neonate primate's reaction."

We visited Dr. Denny-Brown's infant rhesus that was born with cerebral palsy. "You can see that it manages very well, considering," said the doctor. The little creature's hands and feet were crippled in a special manner, with fingers and toes clenched, but it moved about its cage nimbly, nevertheless.

"It's the only known case of cerebral palsy in a monkey," said the doctor, "but that doesn't mean, of course, that there haven't been others, only that an animal born like that in the wild wouldn't survive very long."

In the biochemistry laboratory they were working on phenylketonuria, one of the diseases that cause mental retardation in children. Though they are finding most success with the rhesus as a model, New England researchers are screening other species, hoping to widen the area of inquiry.

Dr. Mathiah D. Daniel from Ceylon told me something about the work of his department, the Division of Microbiology.

"We've found several new viruses in South American monkeys," said Dr. Daniel, "and one of them, at least, promises to be important. Three were found in the squirrel monkey: one is herpes T; the second, herpes virus saimiri; and the third, which hasn't been named yet, was isolated from the squirrel monkey heart."

I recognized the word *herpes*. Like most of us I knew that herpes simplex is the organism responsible for fever blisters, and I had heard that it can be fatal to babies. Moreover, one should avoid getting any of the blister water in one's eye as it can cause blindness. Then there is herpes zoster, more commonly known as shingles—very nasty and painful. Monkeys, evidently, have even more trials with viruses than we do.

"We've also lately found a new herpes virus in a marmoset, but we haven't had time to study it very much yet," said Dr. Daniel. "The owl monkey has several viruses, some of which are herpes and others still unknown—five or six have been isolated to date. Of all this lot, the most important seems likely to be the herpes virus saimiri, which was isolated here by our chairman, Dr. Luis V. Melendez, from a kidney-cell culture. This virus is a latent agent, that is, it seems not to affect its host at all, but"—Dr. Daniel spoke slowly and impressively at this point—*"it causes a fatal disease in the owl monkey and the marmoset*. We don't know as yet if owl monkeys and marmosets in the wild are affected by this organism which is carried by their close neighbor: we do know that in the laboratory it's fatal to them. The disease is characterized as a malignant lymphoma, or cancer, and is the first instance in which we have a primate virus which when inoculated in another primate species has induced cancer. Some of our recent experiments seem to suggest a leukemic picture as well, but we can't say this with certainty: it's still under observation.

"The T virus, too, causes a fatal disease in both marmosets and owl monkeys, but it isn't cancer. It is, rather, similar to the encephalitis caused by herpes simplex in infants. The simplex and T viruses have no apparent relationship with each other except that both are viruses. Yet the lesions they cause are similar. We have been able to produce a vaccine against the T virus, to protect other animals against infection.

"Owl monkeys can also be fatally affected, in both experimental and natural infection, by the herpes simplex virus: and we've produced a vaccine against that, too. All in all, one can say we're spending our time getting a clearer understanding of induced cancer in these species, and in characterizing all the new viruses that have been isolated from New World primates."

Dr. Felix G. Garcia from Havana, trained as a veterinarian before he came to the United States, specializes in clinical laboratory medicine, but what he offered to show me was something other than medicine—a very rare specimen, the surviving third of marmoset triplets. Marmosets, those tiny primates that do not seem monkeylike until one looks closely at their faces, have a faculty unusual in the primate world: they nearly always produce twin offspring—and identical twins, at that. Dr. Garcia's prize marmoset mother, however, went the others one better by producing triplets. As is often the case in multiple births the infants were undersized, possibly premature. Two of them soon died, but the other survived and was living in an incubator in the center nursery.

I followed Dr. Garcia through this room, where two white-coated girls were feeding rhesus infants out of doll-sized nursing bottles, to the far wall where the incubator was. I have said that the marmoset babies were undersized. Keeping in mind that this particular species of marmoset—the cotton-top—is very small even at its largest, you can possibly imagine how very tiny the infant was, but I wasn't prepared for it, nevertheless.

When Dr. Garcia lifted the lid, I blinked. All I could see for a moment was an expanse of white mattress with a small detail in one corner, something like a hairy black marble. Dr. Garcia very gently lifted the sheet that covered the rest of the animal, exposing its tiny, perfect monkey body, complete with tail. In all it was no bigger than a mouse. The head moved; the minuscule eyes opened and looked at us, then closed; the baby opened its mouth and squeaked. Softly Dr. Garcia covered it again and closed the incubator lid. We tiptoed away.

7

The Free Floater

The primate center at Davis, California, is special. Each center, of course, is not exactly like all the others, but Davis is special to a higher degree, and its difference is implicit in its name—the National Center for Primate Biology, University of California.

In 1959, when the first six research centers were authorized by Congress and their lines of research were blocked out, somebody commented that one important function had still to be catered for, since none of these centers could be expected to take care of animal husbandry in a big way. What was needed, in the words of the Davis brochure, was "a facility dedicated to procurement, housing, breeding and manipulating of diverse species, and to interdisciplinary investigations of the broad biologic features of these species." Regional Primate Research Center language has a tendency to use mouth-filling words like that, but one soon catches on. At least, I did.

"It was to be a conditioning center," said the present director, Dr. Robert E. Stowell, more tersely. "The initial grant was voted in '62 and it opened in the fall of '65, but it took a while to get off the ground. At that time I was elsewhere on the Davis campus."

Davis, not far from Sacramento in northern California, appeared to have everything needed for animal husbandry, even the husbandry of such way-out animals as monkeys and apes—a mild climate in which these exotic imports could live out of doors, plenty of space for a project that would call for at least three hundred acres, and men experienced in methods of animal production, for the University of California at Davis has five colleges of agriculture and veterinary medicine. The brochure lists five main objectives: (1) to develop methods for importing, conditioning, housing, handling, husbandry, breeding, and disease control of "substantial numbers of primate species; (2) to study systematically the biology of the various species, presumably while there are still good large groups at hand, before they are scattered among the other centers—these studies to be of all sorts, "ranging from morphological to behavioral in scope"; (3) "to exploit the unique biologic characteristics of various subjects for solution of problems of concern to the human primates"—though it might be argued that this is just what all the other centers too are doing; (4) to provide facilities and test objects for investigators from national and international scientific communities which could not be carried out in their home institutions; and (5) to furnish unique experimental subjects to workers outside the center. Such might include special breeding stock or animals with unusual metabolic qualities, or age.

Of course, all or nearly all of the other six centers have breeding programs for the one or two species used extensively in their particular research problems, but the scope of the national center's project outdoes any of these. At Davis the officials make a conscious effort to think of the primates flooding in from foreign parts in terms of herds, as if they were cattle or sheep. At first, I confess, I found it difficult to contemplate monkeys as breeding stock, but two days at Davis got me over this hurdle, and I slipped into the local way of thinking.

The associate director, Dr. Lloyd J. Neurauter, has been associated with the primate centers almost from their beginning, having spent some time in this work at the NIH headquarters at Bethesda. On the way from the airport he told me something of the Davis Center's troubled early days, when disastrous epidemics attacked large numbers of the animals and for a while threatened to wipe out the entire subhuman primate population.

"It was pretty bad," he said. "Bad enough to make you wonder if it's possible to keep large numbers of exotic animals at all. But of course it's possible, once you know what dangers to look out for, and how to take precautions. A lot of our work is testing new animals that ordinary quarantine methods may fail to examine thoroughly. Old World primates have a lot of enteric infection and TB, of course: we test all our animals regularly for TB, and test the humans who work with them as well. Pneumonia's another of the common killers. And then there are the parasites: sometimes I think we'll never learn everything we should about those. Two people at the center do nothing but study them.

"Being a veterinarian at a primate center is a very different job from that of the man who takes care of domestic pets and farm animals. Laboratory workers here are also concerned with zoonoses, the diseases that humans can catch from subhuman primates. There's been all that work on viruses since the Marburg deaths. We're very careful nowadays. We weren't ever careless—I didn't mean that—but there were things we didn't know, and are learning about." He slowed the car a little. "There's Davis over there. You'll see the primate center in a few minutes. . . . There. What does it remind you of?"

I replied promptly: "It looks like an African village."

It did, though I was uncertain about the material on the roofs of the huts. It should have been thatch, and it wasn't.

"That's a new one," said Dr. Neurauter, not displeased. "I thought you might spot them for what they are. They're corncribs from Iowa. Of course, they're not as high as you might be used to seeing——"

"I don't think I ever saw corncribs in Iowa," I admitted.

"Well then, that explains it," he said. "To be precise, they're *half* corncribs. It was our idea. We were wondering what kind of housing we could use for so many animals out of doors: cages cost, and we're trying to hold expenses down. One day in Iowa we saw some corncribs, and it just came to us: We ordered a few right away to try out. Because they're a lot higher than we need, we cut them in half—we order twice the number of tops—and that gave us two enclosures for every crib. I'll show you how it works out later on."

We got out in the driveway of a large one-story building that, Dr. Neurauter said, was made of prefabricated materials. "It didn't cost anything like as much as those other buildings you've been looking at," he said. "We don't have the same weather problems, you see. Oh, it rains all right here—we have as definite a rainy season as they have in the Congo, but cold snaps are the exception. When we get one, I admit we're pretty busy running around improving the shelters. A freezing wind even at thirty-five degrees can be much colder than it sounds, especially when you have rain at the same time. Ever hear of the chill factor? Just the same, it's an uncommon kind of a problem and didn't justify a building in stone or brick. Here on our right is a field where we've been growing a lot of our own food, alfalfa and such; later we may have to build on it. Come on in."

I paused in the entrance hall to look at a display labelled: "Embryos and fetuses of macaque monkeys. Normal length of gestation about 24 weeks."

"We have displays like this for the benefit of the public," said Dr. Neurauter. "We encourage people to come and look at us: we have guided tours for classes and all that, and I arrange for little talks beforehand so they'll know something about what they're seeing. Kids today are surprisingly knowledgeable, you know—they're really interested. We have stuff for any number of displays, since it's one of our functions to exchange material with scientists everywhere in the country. Nothing elaborate, you see—that's a burlap screen in the background."

The director, Dr. Stowell, received me in his office-cum-library. He said, "This is the largest primate colony in the country if not in the world; and it's probably going to be larger. At the moment we have more than two thousand animals, of nine different species. There are breeding colonies of eight species, which presumably is just the beginning. Some day we may have as many as twenty-five thousand animals, but that's a very long-term project. By that time we'll probably be using all of our three hundred acres.

"You've read the outline of the program, I suppose. Our main objects are husbandry, reproduction and baseline studies such as blood chemistry norms, hemoglobin factors, parasites, bacterial diseases, reproductive physiology, menstrual cycles. . . . In addition, we provide animals for research, and visiting scientists can work here. We have quite a few now: a young veterinarian doctor, Kaye Smith, is studying an atypical tubercle bacillus that seems relatively new: she's trying to find out where the animal got it. Possibly it came from domestic livestock here at Davis. This must be worked out.

"We have the only breeding colony of Indian langurs that I know of, about fifty animals: they're doing well and have twelve young ones born here. Dr. K. F. Meyer of San Francisco imported them to study plague, and handed them over to us to keep. Dr. Kilham has brought baboons from West Africa to study a feature peculiar to these particular animals: they are susceptible to light flashes, and go through what looks very like epileptic seizures when exposed. There are forty baboons in that troop, and very likely they'll increase.

"Two doctors from the NIH are studying slow-acting viruses which may produce disease after months or years. Perhaps such viruses are responsible for kuru—a slow virus disease of the nervous system—certain neurologic diseases, and Parkinson's disease. They're using a chimpanzee for the kuru tests, as well as squirrel monkeys.

"In husbandry we investigate caging, housing, and care, trying to find the least expensive methods for these. Neurauter's told you about our corncribs, which take an average of fifty animals each, but we also try other types of enclosures. We study how far apart these cages should be kept to avoid the spread of disease. We have to experiment with various types of base material, concrete, and so on. We must also explore various devices for shelter and entertainment. Monkeys are like people: they're not all alike. Some need one type of shelter, some another, and the same is true of entertainment. We've refurbished the indoor cages and replaced a lot of the older facilities, the operating room and so on.

"We're also rebuilding the staff. When Leon Schmidt left two years ago, a lot of people left with him. The University of California came to our rescue when everything looked at its darkest: they advanced funds which we can pay back in several years. We get money from other sources, too—research agencies that ask us to raise and keep animals for them, that kind of thing. On the campus we have a program in the medical school financed by the NIH. That's where I was initially—the medical school, where I was chairman of the Department of Pathology. I'm still a professor there."

"A lot of questions have to be answered when you try to domesticate wild animals," said Dr. Neurauter, "but I don't see why it can't be accomplished, in controlled conditions. . . . We should go in for controlled breeding, the way people do with dogs and pigs and the others. We need to identify the monkey as the ideal animal for research, but first we should decide on the species that would answer our purposes best—the all-purpose primate. If we could only miniaturize the chimpanzee, that would be the ideal animal."

I looked at him wonderingly. No, he meant it. I thought of the time it would take to miniaturize a chimpanzee, and felt a little dizzy. Poodles, yes: chimps, no, I felt.

"There's no question but that they're valuable, even as they are, to space flight," he said dreamily.

We entered the center lecture room where a high school class was waiting to be indoctrinated. I took my seat with the youngsters and listened. Dr. Neurauter's talk impressed me: it was matter-of-fact and not patronizing. "We all recognize the fact that primates are closely related to us," he began easily, "more closely than the armadillo, for instance."

Afterward we walked through the corncrib village and looked at the animals before returning to the buildings and laboratories, where one doctor said to me earnestly, "Problems with the primate in captivity are *different*. They have hands. They're manipulative, and they're clever. Their relationships and social structures are very important in their lives, which is why we seem to spend so much time on behavior. I know the world outside thinks we do too much on behavior—somebody said the other day that primatology is overstocked with Harlows, but it's not true: we aren't wasting time. You can't handle an animal in captivity unless you know how he behaves in the wild, and monkeys are too intelligent to fool. Usually they catch on to things quicker than the animal handlers."

In a final conversation, Dr. Stowell said, "Sometimes I wonder why we don't concentrate on three species for research—for instance, the rhesus, the baboon, and the bonnet. We could really raise them on ranches then, and give the orang and all the others back to the conservationists. On second thought, no substitute would do for the chimpanzee, I suppose, and then there are all these new things they're finding out about the New World monkeys. . . ."

He sighed, like one relinquishing a dream. "The others for the most part are arboreal and hard to keep. Of course, in a way we too are conservationists, but that's not our main job. I'm tempted to say, Let's leave the dusky langur alone!"

The Delta Regional Primate Research Center is connected with Tulane University in New Orleans, but one would never guess it, for it is some forty miles from the campus, across the causeway, near the town of Covington. Delta does a good deal of work on infectious diseases, and it was considered wisest to put it where it is. Deep in the country, set in some of the prettiest landscape in the United States, it looks at first like an outsize country house. Indoors, however, the impression disappears—they are a busy crew.

The last time I visited the office of the director, Dr. Arthur J. Riopelle, I saw something new in the way of decoration—a large ape doll, obviously a replica of a famous albino lowland gorilla in the zoo at Barcelona. Copite de Nieve, or Little Snowflake, is the albino's unlikely name, which probably didn't seem so unlikely back in 1966, when Copi was an appealing two-year-old in the Rio Muni forest of Spanish Equatorial Guinea. He was captured when his mother, an ordinary black animal, was killed by a hunter. He has typical albino coloring—white hair, pink skin, and blue eyes—and the doll in Dr. Riopelle's office had the same. I assumed, correctly, it was a Barcelona toy.

"Who chewed off its nose?" I asked.

Dr. Riopelle said, "Copi himself did that. A couple of us go over to Barcelona every year to see him, measure him and all that, and this time we tried the doll on him. He's got a cage mate, an ordinary black female gorilla slightly younger than himself, but the fascinating thing about him when we handed him this doll was that he reacted to it in quite a different way than he does with his playmate. He recognized it as something like himself. He grabbed it right away and even reacted sexually. Naturally he's seen himself more than once in a mirror, so I suppose it's not as remarkable as it seems. Just the same. . . ."

We discussed the latest developments at Delta: Dr. Oscar Felsenfeld, he told me, working on enteric diseases and generalized infections, had been applying the knowledge he gained to that scourge of most parents, infant diarrhea.

"He's had a lot of experience with enteric diseases and tropical medicine generally in the Far East," said Dr. Riopelle. "He used to be director of the SEATO medical laboratory. One of the things he's particularly interested in is cholera. We can't be too careful of that, what with stepped-up transport and everything; it could migrate through the Middle East and happen here. His wife, a Thai girl, is a virologist: she's working on infectious hepatitis. Then there's Tom Orihel, who's engaged in parasitological studies, especially filariasis. Did you know that can be tied in with heartworm in dogs?"

They had six wild-born chimpanzees, he said, all having been inoculated with hepatitis material and all being studied. Counting the healthy animals, they had sixty to sixty-five chimps. Yerkes and Delta are the only two centers with equipment that makes it possible to keep so many great apes as well as the more usual monkeys.

Dr. Riopelle continued: "Kenneth Brizzee is working on the effects of low-level radiation in the susceptibility of brain cells to further radiation; we're hoping this will lead to the development of more radioprotective drugs. And a young couple are here from Taiwan: they're studying how blood components react in response to ageing and disease.

"We've got a group in reproductive biology busy on several projects. One is to find the suitability or otherwise of a small African monkey, the talapoin, for research of this sort. They're trying out a uterine device on that monkey—perhaps you know that the trouble with these devices is that they might slip out. The hope is that if they're coated with a certain hormone, they won't slip. Harold Spies is working on neural control of reproduction and reproductive behavior, and L. E. Franklin is investigating the morphological changes that occur when the sperm first comes into contact with the ovum. This is a very critical period in the life-span of an individual. We don't know to what extent developmental disorders start at this moment, when something might easily go wrong. Usually if things do go wrong, you don't get a successful take—but if you do, the resulting animal might be badly affected. Also, if you learn enough about the process, you might develop a different contraceptive technique which doesn't involve steroids: a lot of people have doubts about steroids because they aren't sure of the long-term effects on women.

"We don't know nearly enough about reproduction in primates. When I was at Orange Park we had a female chimp about forty years old who had gone ten years without seeing a male animal close up. She was still cycling, so one day I tried an experiment: I opened the cage door so a male could get in, and in ten seconds that chimp was pregnant. I almost lost my head over that," he added thoughtfully. "And it ties in with something else we ought to do a lot more of—studies of old age. At Orange Park we were actually criticized for doing the small amount we did: it was said that we were running a hotel for ancient apes. Well, why not? The only data we've got on old age in primates is from those ancient apes: we had continuity, which you don't get nowadays. Most grants are for short periods—three to five years—with the result that instead of good solid research jobs you end up with any number of short-term studies in neat little packages that are small good to anybody.

"Charles M. Rogers and I once did a complicated study on the performance of aged apes; we found out a lot, but that was in the days of Orange Park. In our original chimp study here we hoped to follow animals through their whole life-span, but then came the financial squeeze. Animals now are systematically set aside when the project is discontinued every five years, perhaps only to be cranked up again in another five years—starting all over again. I guess an old-age study or a life-span study, if it's done right, takes an endowment kind of commitment."

On my first visit to Delta I had been taken by Dr. Emil W. Menzel, Jr., research associate and associate professor of psychology, to watch a group of five-year-old chimpanzees. Every day they played for a while in a field cage about an acre in size, in which stood trees—which soon, under their treatment, became stripped, dead trunks—and an observation platform. The chimps were six in number, and though they appeared to me to be a single group, they were really, Dr. Menzel explained, two groups of three and two, and one free floater. The trio were Shadow, Belle, and Bandit; the pair were Libi and Bido; and the free floater, an expression that will be explained in a moment, was Polly.

That day Dr. Menzel, an assistant, and I stood on the observation platform and watched the chimps run out as soon as the door to their cage was released. The assistant held a timer that gave a signal at fixed intervals. Each time we heard the signal Dr. Menzel dictated a commentary on what the animals were doing. Two might be walking together, "tandem," arms around each other's waists or leaning on each other's shoulders; or one might wrestle with another or play follow-my-leader.

The point was that no matter what they did they never got very far apart from each other. Dr. Menzel took two groups, or subgroups, each of which had much the same history—in each, that is, the members had been living together for a considerable time, varying from three months to two and a half years. Neither group, however, had had any experience with the other until the whole lot were put together in a one-acre field cage and housed there permanently.

Dr. Menzel wrote, "The basic problem of the research was: What constitutes a 'social group,' and what holds an entire aggregate of chimps together in space across time, or causes it to separate into smaller parties? . . . Initial tests and continued base-line data on patterns of group travel and social interaction, clearly showed that the subjects quickly came to travel together as a single aggregate, and to interact freely with any other animal. However, any given subject's primary associations were with members of the original 'natal groups.'

"At night the subgroups invariably separated to sleep in different locations. Shadow, Bandit, and Belle, who had been raised together since infancy, all slept together in one cage; Libi and Bido, who had been raised as a pair, slept in another hutch; and Polly, who had spent several months before the experiment living with Libi and Bido, but before that had been rotated from one housing situation to another, living alternately alone and with different companions, usually slept alone in a third cage. If disturbed at night, however, Polly often left her cage to join one of the other parties.

"During the day, even after 1 year, subjects seldom ranged farther than 100 ft. from an old companion, and when the entire aggregate spread out over a substantial area, with no one in direct physical contact, old companions usually followed each other and stayed closer to each other than to any other subject. The typical pattern of travel, especially through open areas or when subjects were cautious, was 'tandem walking,' with old companions holding each other about the waist and walking in a chain—usually in the same orders within the chain. Most rough play and all fighting was across subgroups. When fights developed between a pair of subjects, all animals usually came running to the scene, and each then 'took sides' with the member of his subgroup. The only subject whose affiliations wavered or fluctuated over the period of a year was Polly."

Oh, yes, Polly, the free floater. On the day of my visit Dr. Menzel's dictation went something like this:

"Bandit climbing, followed by Shadow and Belle. Libi and Bido tandem. Polly about thirty feet to the rear.

"Bandit and Belle tandem, Shadow behind. Libi and Bido climbing. Polly thirty feet to the left.

"Libi and Bido tandem, Bandit chasing Polly. . . ."

It went on for about an hour before I said, "Dr. Menzel, I think I'd better go now. I know you'll think it silly of me, but I can't help identifying with Polly."

A year and a half later, I learned that Dr. Menzel had been working on new exercises for his six chimpanzees, and had written a paper about it.

"One of the basic reasons that group members separate is presumably to reach spatially separated ecological goals," he wrote. "A series of experiments was therefore conducted using a wide range of usually attractive objects—novel toys, household articles, trees, people, etc. All 6 subjects were confined in a release cage before a trial began, and 2 or more objects were placed out in the field at a distance from this point. Then the cage door was opened and subjects were free to respond."

He discovered that the six animals rarely split into parties to approach two objects at once, even when the objects were only fifty feet apart. Instead, they worked as a single unit, going first to one of the objects and then to the other. Later, getting used to the game, they took such objects as they could carry to "a common preferred location," where they did begin to split up to some extent. "Old companions often shared or exchanged toys, while driving others away." Even so, unless he was chased, no subject went more than twenty feet from the others to hang on to an object, and "this close aggregation of animals and desirable objects occasionally led to fights between entire sub-groups."

They tried the chimpanzees with food prizes as well as toys. It has been said by other chimpanzee watchers that food seems to be "a uniquely important class of objects for increasing population dispersions," but Menzel noticed that his "well-fed juveniles," once they had reached their goal, fought harder for toys than for food. "Very few fights occurred over food. . . . Much more common responses to losing a race were clinging to the victor, begging for food, initiating social play, or grabbing the food stake and wrestling, mounting, and thrusting on it."

Again, even when the group split it was only for a minute. As soon as they had grabbed the prizes they got together again. "Apparently chimp groups do not travel together . . . merely because they happen to be going to the same goals; there are also forces of some sort that pull them back together once they chance to separate."

In later tests, when the apes were older, their habits changed to some extent. Six supplies of food were exhibited in different locations, the situation always visible to all the group. Sometimes four lots of food were put on one side and two on the other, sometimes five to one, and so on. Except when all the food stakes were on one side, the group almost always split—but always with a neat division, more chimps going to the larger collection. They seemed able to sum up the situation literally at a glance and as one chimp. "About the only discrepancy in these data," says Dr. Menzel, "was produced by a single chimp, Bandit, who rarely competed for distant food, but frequently begged or stole it from the successful." Polly seems to have become integrated.

Later, after yet another lot of tests, Dr. Menzel delivered a report as a lecture in Boston. I read the paper before going to Delta for the last time. He began, "My general problem is, where will a chimpanzee be next, and why does he go there rather than elsewhere? The preliminary answer to this is simple: A chimpanzee will generally be where his close companions are. Indeed, if we knew the minute-to-minute location of a social group's center, we could account for at least 90 percent of any given individual's movements over the course of a day. The question that immediately arises is, Where will the 'group as a whole' be next? What controls its location, its general direction of travel, and the moment-to-moment scatter of individuals about the group center?"

To answer these questions he devised a game. The chimps were locked up where they couldn't see the field, while Dr. Menzel and his associate hid food under a pile of leaves and grass or behind a tree. Then Dr. Menzel took out one of the animals, which was nominated "leader" for the purposes of the game, and carried it to where the food was hidden. He showed the food to the chimp but didn't let him have any, then carried him back and reunited him with the group before releasing the whole lot of them.

The leader tried to persuade the others to follow him to the food. If nobody followed him, he got very upset. "He would go from one follower to the next, grimacing, tapping him on the shoulder, starting off tentatively and then stopping to glance backward, or (in the extreme case) screaming, grabbing a preferred companion and physically dragging him in the direction of the food. Once the followers followed, however, the leader soon turned his companion loose and ran—and this usually produced a pack race. All this, of course, suggests that group cohesion was strong and the 'leader' was as dependent upon the group for getting to the food as they were dependent on him in knowing precisely where to go."

Next a variation was tried. The men showed the leader the food secretly, as they had done before, but after returning him to the group, they released him again without the others, keeping them in the cage. This gave him a chance to go and get the whole lot of food for himself, but did he take that chance? No. He whimpered, defecated, begged, and even tried to open the cage door himself to release his friends. When Bandit was leader, he had a tantrum at such times, screaming and rolling on the ground, then running and clinging to a tree. (Three times, it is true, when Belle was leader she ran and got the food for herself, but only when it was hidden very near to the cage.) Once the door of the cage was opened, the leader's behavior changed rapidly. He "ran to a follower, screamed and embraced him, and within five seconds was off and running for the food."

The men were much surprised to see that two leaders, each with knowledge of a different goal, were able somehow to communicate to each other the relative value of these food caches, so that the whole mob would go first to the bigger lot and then to the lesser. Dr. Menzel asked himself how this was brought about. What did each leader do that changed the behavior of the other leader and of the followers? He had seen many instances of behavior on the part of the leaders that would interest those who study primate communication—"For example, glancing back and forth between the goal and the followers, arm signals, 'presenting' the back to get another animal to walk in tandem, whimpering, and tapping a follower's shoulder or tugging his arm, occurred fairly frequently." But these signals were seldom very effective, and they didn't happen often enough to explain all the data. Dr. Menzel observed that the best strategy a leader could use to attract followers was to get up and move off independently in a consistent direction. He thought that the most likely explanation for the group's ability to go to the right place was not some special form of chimp language, but the general tendency of any individual to run faster, go in a straighter line, and ignore other goals such as companionship—or, for that matter, to pull harder, whimper louder, or gesticulate more vigorously. There is also, he said, the ability of chimps, or most mammalian species for that matter, to use another animal's behavior as a cue that something is 'out there' in the environment.

Having read Dr. Menzel's papers since earlier visits to the center, I was eager to see the chimps again, so he took me to the field. As we approached the fence, the apes came tumbling up from all directions to get a closer look. Then, as if satisfied, they returned to their former pursuits.

"They *have* grown," I commented. "They don't act at all the same."

"Far more independent now," agreed Dr. Menzel. The chimps were, indeed, all over the place, each one behaving with such assurance that it was hard to believe he had ever clung constantly to a companion, or made only tentative, fearful steps toward the center of the area. One of them—Bido, I think Dr. Menzel said it was—picked up a small plank, balanced it in the air, and brought it to the wire fence near us, where he began trying, with great patience, to wedge one end into an opening.

"Still, they don't readily separate very great distances, you'll notice," the doctor said.

Ahead of us, one of the dead trees showed an interesting pattern to which he called my attention—a pair of muddy footprints that went straight up the side of it. "A stranger would have nightmares trying to figure that out," he said.

Bido had succeeded in wedging the end of the plank into the fence: now he had a rising ramp that led, evidently, nowhere except straight through the fence. Instead he bounced on it and did a back flip, and I clapped.

"They've begun doing something new," said Dr. Menzel. "You remember what Jane Goodall saw chimps doing in the wild?" He was referring to a young woman who had lived for extended periods of time in East Africa near one particular band of wild chimpanzees, watching their behavior and finding out a good deal that was new to anthropologists. "They trimmed twigs and stuck them into termite nests and all that, and the termites collected on the twigs and the chimps ate them," Dr. Menzel reminded me.

"Yes, I remember," I said.

"Well," said the doctor, "the chimps are doing that here. Not that we have any termites, of course, and the ants are inedible: the chimps never eat them. Just the same, now and then I see a twig sticking up in an ant hole."

8

In the Wild

When Herodotus came back from Egypt he brought with him wonderful stories of people he had seen at a king's court—fully adult men no bigger than children. Most scholars believe that these were pygmies from the tribe that still inhabits part of the northeast Congo and western Ruanda: pygmy country is not too far from the Nile for slave traders of Herodotus' time, about 450 B.C., to have made their way there and collected a few of the inhabitants. The traders might well have created a demand for the pygmies as curiosities. But it seems possible—to me, at any rate—that what the Father of History encountered were not specimens of men at all, but young, manageable chimpanzees.

If Herodotus mistook such anthropoid apes for wild, hairy humans, he would not be the first or the last to do so. The chimpanzee so much resembles us, and adapts itself so easily to our way of life, even to wearing clothes, that many people, including this writer, find it almost impossible to relegate it to the outer world of nonhuman animals. At Tange in the Ituri, where I lived for a time years ago, the Africans referred to pygmies by a special name: "Is he a pygmy or a man?" I heard one villager ask of another, indicating a somewhat undersized stranger. And the same, in reverse, applied when they spoke of chimpanzees. Once when I demanded angrily, "Who pulled down that tent?" the cook replied, "No other person than Chimpo."

At many other points as well, the line of demarcation is blurred. Pygmies do their thing a bit better than chimpanzees—that is all. They build temporary sleeping places, and so do chimpanzees, but there are differences: A pygmy house stands on the ground and is complete, with a roof made, like the walls, of plantain leaves, whereas a chimpanzee nest is usually up in a tree, and it has no roof. Pygmies can talk our language when they try to. It is a moot point whether chimps can talk at all: certainly very few have ever tried. Pygmies are accomplished hunters, stalking and bringing down elephants, but those few chimpanzees that are known to catch animals go after very small game, in a primitive slapdash manner at that. There are lots of such differences. Nevertheless, if Herodotus made a mistake, we should not wonder at it, even though in the intervening centuries we have become fairly sophisticated about apes.

Probably the first person to mention these animals (after Herodotus, that is) was the English sailor Andrew Battell, whose account was given to Samuel Purchas in 1613 and subsequently published in the book "Purchas His Pilgrimes." Battell had been kept prisoner by the Portuguese for years in Angola. He tells of the Angolan province of Mayombe on the coast, "where the woods are so covered with baboons, monkeys, apes and parrots that it will fear any man to travel in them alone." Among these creatures were two monsters, common there and very dangerous—one enormous monster called Pongo, obviously a gorilla, and a slightly lesser one, Engeco, which must have been a chimpanzee. The word *chimpanzee* itself derives from an Angolan dialect, and means "mock man."

In 1641 an Amsterdam physician and anatomist, Claes Pieterszoon Tulp, published a work, *Medical Observations,* illustrated by at least two anatomical drawings—of the narwhal and the chimpanzee. His model for the latter was an animal owned by the Prince of Orange, and it was possibly the first living specimen ever to reach Europe. Afterward more chimpanzees arrived; not many at first, and they did not survive more than a season, but as long as they lasted they attracted fascinated attention. If for years few people evinced curiosity as to the chimpanzee's true nature, this in a way was the ape's own fault. To watch the animal make himself at home in houses, to teach him to behave like a human being, was so amusing that it distracted the watchers and led them away from the spirit of inquiry. Men made copious notes of chimpanzee behavior, but they were all of how chimps acted in our world, not in their own. Even today performing chimpanzees are a sure-fire attraction in the circus, on stage, in TV shows, and even as night-club acts: people never seem to tire of watching the animals prove that they can eat at table, ride tricycles, and rollerskate. The only information past generations gleaned about chimpanzee behavior—with the honorable exception of Richard L. Garner, who went to Africa in 1900 to watch primates—came out of stories told by Africans to gullible travelers. Then, early in the 1920's, Dr. Robert Mearns Yerkes at last took action. In addition to establishing a breeding colony of chimpanzees at Orange Park, he visualized field trips where the species could be studied in its natural habitat.

There was potential competition for his venture. When scientists at Pastoria—the laboratory complex at Kindia in French Guinea—needed a chimpanzee all they had to do was step outdoors and catch one. Yerkes felt he ought to have a look at the station, though as he explains in his foreword to a monograph by his assistant Henry W. Nissen, "A Field Study of Chimpanzees," published in 1931, he wasn't really worried. The French scientists weren't interested in psychobiology, but in medicine. He wanted to see their equipment and methods, and he still hoped to do fieldwork, but for some years after Pastoria began operating, he was unable to get away from Orange Park. Then in 1927 he heard that the Russians, too, were starting a primate center, at Sukhumi on the Black Sea, and he resolved to visit both places. In 1929, in the summer, Yerkes got to Kindia at last.

He had the best possible credentials. The Pasteur authorities were anxious for financial help from America for their Kindia project, and Yerkes had been requested by officials of the Rockefeller Foundation to take a good look at the setup and give them a written report on his return. The French cordially agreed to his proposal that someone from the Yale Laboratories of Comparative Psychobiology come to French Guinea and make a systematic study of the life of the wild chimpanzee, also to bring back to America a group of young chimps for Orange Park. In December, therefore, after Yerkes came home, Henry Nissen set sail in his turn for Conakry. There he spent two and a half months on fieldwork—not much considering it was the first expedition of its kind, but he was supposed to make a preliminary survey only.

The literature on chimpanzees is full of valiant attempts to sort out various subspecies. Early observers in the very young study of primatology made divisions based on all sorts of features—the color of the animal's face (for some chimpanzee faces are black, some pink, some freckled); the shade of his hair, which sometimes is reddish rather than black; the thickness of hair growth; and the baldness or lack of it on the head. But these variations often prove to be individual, not specific, and today's primatologists have more or less agreed to recognize only four, or possibly three, main types: *Pan troglodytes troglodytes, Pan t. verus, Pan t. schweinfurthii,* and—perhaps—*Pan t. paniscus*. The last named may in fact be a separate species, in which case it would be *Pan paniscus*. It is the pygmy chimpanzee. Nissen's animals were *Pan t. verus*.

He selected for the survey an area about twelve and a half miles in diameter, the center about nineteen miles from Kindia. The chimpanzees of the region had often been raided by men collecting for Pastoria, but Nissen reasoned that Africans had long hunted them in any case, for meat, and he could only hope the local animals were not too badly demoralized. The fieldwork began in February 1930 in the middle of the dry season, and he escaped with less than twelve hours of rain the whole time he was there. The heat was "oppressive and fatiguing"; one needed constant protection against sunstroke, he wrote, but the natives didn't seem to mind it—not as much, apparently, as the chimps did. It was a hilly country with narrow streams. Patches of lush vegetation were interspersed with grass that grew up to five feet and burned dry and brittle in the summer.

There were no guides on how to watch chimpanzees, so Nissen started out in the conventional way of the naturalist, by constructing a few blinds. He wasted two days sitting in them, for he never saw an ape the whole time. He concluded that they were too alert to be fooled in this way. Also, using a blind depends on the habits of one's object, the daily routine of the animal, and as Nissen was to discover, there was nothing regular about chimpanzees. They might congregate in one clearing for three days, eating fruit that had ripened there; then for no ostensible reason, they would move on to another feeding ground. The easiest way to locate them was to listen, since on the road the animals made a lot of noise, yelling and drumming. The drumming they did by hitting hollow trunks or tree buttresses hard and rhythmically—it often sounded, he said, like tomtoms, and could be heard a very long way off. Nissen did better, he found, when he left his Africans behind and tried to get close to the apes, at least close enough to see them with binoculars. Once or twice he succeeded in approaching to within fifteen or twenty feet, but most of his observations were made from at least fifty feet away. The most satisfactory method was to send out scouts in the evening to discover where a group had gone to bed, and next morning he would be there before they woke up, to follow them as long as he could. Twice he stayed all night near a sleeping party.

By examining feces, he learned a good deal about chimpanzee diet. They ate a wide variety of leaves and fruits; he never saw them eat anything else. He was able to confirm that chimpanzees were highly socialized and almost invariably lived in bands or groups which some people called "families." Nissen would not call them families, arguing reasonably that he had no way of knowing what relation they bore to each other. The smallest group he saw consisted of four members, the largest of fourteen, but on two occasions he saw groups together in a sort of mixer, or combination party, when they numbered sixteen and eighteen respectively.

"It is gospel among the natives," he wrote, "that a chimpanzee group consists of the male leader, his wives and his children. The family of the French Guinea negro is thus constituted, which makes it probable that his observations of the apes are strongly tainted by anthropomorphic interpretation." In fact, he continued, six times he had seen groups containing not one but two mature males, and though they might represent father-son relationships, other interpretations beside that of the Old Man system were possible. Perhaps neither male nor female chimpanzee is limited to a single mate, even temporarily, he speculated; and later studies made by others were to prove him right.

At dusk the chimpanzees made nests to sleep in, collecting branches and weaving them together in the trees where they elected to spend the night, filling the shallow depression in the middle of the arrangement with twigs and leaves stamped on and patted until they offered a smooth surface. Members of a group built their nests close together, all in one tree or at least in nearby trees, at a height of from 13 to 105 feet. Nissen took several nests apart, knowing they would not be missed, since as far as he could see no animal used the same nest twice. He found that a typical nest was almost flat, with practically no sides. It was oval in shape, nearly round. One that he examined was twenty by eighteen inches across. Most of the branches that made up the base had been bent backward—probably to give a springy feeling—and the inside was very smooth.

Reading this part of the report put me into reminiscent mood. It was strange, I reflected, that I too should have been dealing with a chimpanzee in Africa in that same year of 1931, though I was on the other side of the continent, in the Belgian Congo. Chimpo was not wild, but she was not captive either, and once she too built a nest, in a tall tree near my front door. The tree was an ironwood and looked like a giant feather duster, with no branches on the way up, but then, high in the air, a burst of them thick with twigs and leaves. I don't know why Chimpo suddenly made that nest, or rather why she had never done it before; perhaps it was because she had always been put to bed in the house when dark fell, and never had the chance.

At any rate the impulse took her at high noon that day. An hour or so later, I was reading on the veranda when I heard a familiar sound, to which at first I paid no attention. Why shouldn't a goat bleat? But my subconscious plucked my elbow and said insistently, "Goats don't fly, goats don't fly," until I put down the book and really listened. Sure enough, the noise was coming from an unusual angle. I went out into the yard to look, and there, peering anxiously down from high in the tree, was a goat standing in the chimpanzee's new nest. It is hard to describe how strange he looked up there. Having carried him up and deposited him, Chimpo was no longer with him: she had become bored and gone away on some other project. It took us all afternoon and three ladders tied together to get the goat down.

Nissen discovered that some of the chimps, especially the larger ones, took a siesta in the middle of the day, in branches or crotches of shady trees or, more often, on the ground. They never used regular night nests for these afternoon naps, but sometimes an ape would pull bush grass down, or use small bushes or little trees to make the ground softer. Nissen called such places daybeds. Some of them were fitted with roofs or sun umbrellas, made of young saplings twisted so that they bent over and shaded the reclining animal. He found six of these elegant arrangements, all near the margins of woods and open fields: in the forest chimpanzees needed no protection from the sun.

As long as you went about your own business, he discovered, chimpanzees didn't mind your being about somewhere. It was only when they thought you were paying attention to them that they ran away. "In one instance I observed chimpanzees feeding in a tree not more than 250 feet from where a dozen native women were noisily washing clothes, talking and laughing," he wrote. "The women paid no attention to the animals and the latter seemed totally unconcerned about the women. But if chimpanzees caught us *looking* at them from much greater distances, they immediately took to flight."

"Although my relatively short stay in the bush of French Guinea resulted in the accumulation of more factual material than I had dared to anticipate, the story of the chimpanzee in his native habitat has only been started," he wrote in conclusion. ". . . The major share of the task remains to be done. That it needs to be done by thoroughly trained observers is obvious. That it is very much worthwhile doing, I think is equally clear."

Yerkes hoped to send a full-dress expedition to Africa soon after the monograph appeared, but the times were against him. First there was the great depression, then the Second World War and its subsequent upheavals. It was not until 1960 that a second scientific observation was made of chimpanzees in the wild, and the man who undertook this task was not a Yerkes scientist, but the Dutch Dr. Adriaan Kortlandt, who went to Africa under the sponsorship of the Institute of National Parks of what was then—as he says—the Belgian Congo, and published the account of his observations two years later. The situation had changed noticeably during the three decades that had passed. In many areas the apes, because of their value as laboratory subjects, had been hunted to extinction, or at least had moved out of the range of observation.

The region selected by Dr. Kortlandt lies in that part of Africa where such chimpanzees as survive belong to another subspecies than those of Nissen's experience: these are *Pan t. schweinfurthii.* He located himself at the edge of a fruit plantation in the eastern Congo. The plantation's Belgian owner did not grudge the chimps the papaws and bananas they took: it amounted to little compared with the full yield. The apes' favorite sleeping place was a wooded hill just at the edge of the clearing: it was protected from human marauders by the local legend that trespassers who harmed the animals would be destroyed by the spirits that lived there. Moreover, the natives believed that chimpanzees could catch any spear thrown at them and throw it back with deadly effect. That plantation was the nearest thing an observer could find to a chimpanzee heaven.

Dr. Kortlandt prepared a number of blinds, and had better luck with his than Nissen had thirty years earlier. One was twenty feet up a giant tree at the bottom of the hill, just where the chimpanzee trail led from the woods into the clearing. Two more were placed at the side of the trail, but the most spectacular was a platform eighty feet off the ground in another giant tree that commanded the whole area in which he was interested, the only drawback being that he had to get settled there before the animals appeared in the morning and stay until they had gone away at twilight. As they spent a lot of preliminary time in the forest before making their first entrance into the clearing, Kortlandt often waited hours for the curtain to rise.

You could hear them coming, however, a long time in advance, yelling and calling. Suddenly they would fall silent, and then the adult males, usually, appeared first. A "broad black face" would peep cautiously through the leaves, looking around before its owner moved altogether into view. Then, one by one, the males emerged, each one looking around and listening as the first had done before stepping out. Even then, at least one of the animals, or possibly two or three, walked upright for a while, looking the land over. Suddenly, as if they had at last decided that they were safe, they broke into what Dr. Kortlandt describes as a wild and deafening display, yelling, rioting, chasing each other, shrieking and screaming. "They stamped the ground with hands and feet and smashed tree trunks with an open hand. Sometimes they pulled down half-grown papaw trees. Occasionally one of them would grab a branch or throw it while running full tilt through the group."

It was different with the females, whom he describes as almost always silent, wary, and shy, especially when they had children with them. They were fussy mothers who seemed to be preoccupied with keeping their young safe. Dr. Kortlandt was surprised to see young chimpanzees far advanced from infancy—as old as four years—still being carried around even for short distances. He was surprised because he had seen many chimpanzee mothers and young in captivity, where the youngsters were far more enterprising and the mothers much more permissive—or perhaps I should put it the other way around.

Apart from being watched so much, however, the little chimpanzees of the eastern Congo had a wonderful life, he thought. By human standards they were pampered, allowed to do whatever they liked. He saw mothers handfeeding quite large adolescents of six or seven years. Children could invite themselves unrebuked to eat with strange mothers and children. At least twice he saw a child run past an adult male and give him a hard smack on the rump without arousing any anger. And yet, in spite of all this indulgence, they weren't spoiled. They never whined or whimpered, and they always obeyed orders promptly. How different, reflected Kortlandt, from the home life of the zoo chimp children he had known! Later he heard from the anthropologist Jane Goodall's mother that among the chimpanzees she had watched in Tanganyika, things were not like that at all. The children were not so cossetted; they became independent far younger; and—or but—they were not nearly so well behaved as those he had watched.

Another feature of Kortlandt's chimpanzees' behavior was their treatment of the elderly. One male animal he knew as Granddad might have been about forty, he surmised. Granddad's silver-haired back was bent, his crown was gray, and his face sagged. He seemed somewhat handicapped: at least, he avoided climbing and seldom participated in the male intimidation displays. Among other primate species such an elder would have been unfortunate, mused Kortlandt, but Granddad was the tyrannical leader of the band, age or no age: all his whims and fancies were indulged, and even the biggest of the senior males sought his company. He acted as a kind of security inspector, too, a superwatchman, always making sure that everything was safe.

In general, the wild chimpanzees compared with zoo animals seemed more lively, interested in everything, more *human*. "This is probably the reason it was so difficult, when I watched them from close by, to shake off the feeling that I was looking at some strange kind of human beings dressed in furs," admits Dr. Kortlandt. "The chimpanzees were unceasingly alert and curious. They seized every opportunity to bring variety into their lives, taking different paths down the hill on different occasions and continually changing their gait and their mode of locomotion. They carefully examined all the objects I laid in their path and even collected some of them. Once I saw a chimpanzee gaze at a particularly beautiful sunset for a full fifteen minutes, watching the changing colors until it became so dark that he had to retire to the forest without stopping to pick a papaw for his evening meal."

Another "human" trait was the way they sometimes seemed to pause and ponder when faced with the necessity of making a decision, such as, Shall I turn left or right? Often at such moments a chimp would scratch himself while he made up his mind. (Laboratory chimps, too, scratch when they are working out problems.) When Dr. Kortlandt was in one of the blinds close to the chimpanzee trail, though the rest of him was hidden, his eyes, necessarily, were visible to close inspection. Now and then a chimpanzee would see them. Sometimes an adult male came straight toward him at such times, and would pause about ten feet off, and look straight into those eyes. When this happened, the chimp never either attacked or ran away, but just stood there for a while, looking thoughtfully into Dr. Kortlandt's eyes and scratching his arms and chest before he wandered off.

Dr. Kortlandt also noticed that except for those moments when they were whooping it up—the intimidation displays, he called them—his chimpanzees were for the most part silent. They communicated with each other by gesture or by "changes in posture or facial expression": If a mother and child were sitting side by side and the mother thought it time to move along, she looked at the child, and it immediately jumped onto her back. Or if it wasn't looking at her just then, she tapped it lightly on the shoulder or arm.

These chimpanzees, he said, definitely used weapons. They brandished or threw branches and clubs. The natives, when they said that the apes could catch spears and throw them back, were not really exaggerating very much.

Soon after Dr. Kortlandt's chimp watching, in 1961, young Jane Goodall began her now famous observations of the chimpanzees (also *Pan t. schweinfurthii*) of the Gombe Stream Chimpanzee Reserve in Tanganyika. Her reports are more detailed than any produced before. To use her own words, the behavior she described "applies specifically to a rather atypical habitat," since the country selected is not the closed rain forest normal for chimpanzees. A mountainous strip runs for ten miles along the eastern shore of Lake Tanganyika and stretches inland to the mountain peaks of the Rift, which attain an altitude in some places of five thousand feet. There are steep valleys and ravines, and though at lower altitudes there is rain forest, one also finds open deciduous woodland higher up, with growths of tall grass. The site was chosen deliberately because it was easier to see the chimpanzees in open country than in the forest.

Another atypical feature of Miss Goodall's study is that she set out deliberately, in order to facilitate proceedings further, to habituate the animals to her presence. It took a long time, but she succeeded fairly well, by giving them extra food and also, when they seemed suspicious, by pretending not to be watching them at all: she would dig a hole or do something else to allay their doubts. There came a time when she could approach quite close without being resented, and after fourteen months the apes carried on their normal activities—feeding, mating, sleeping—when she was only from thirty to fifty feet away. In her estimation the two most interesting behavior patterns observed in the Gombe Stream Reserve were that those chimpanzees ate meat and used tools. Both discoveries came as a surprise to the world of anthropology (but not to me: Chimpo ate meat whenever I did). Jane Goodall's chimps caught and ate young monkeys, young bushpigs, bushbuck, and on one occasion a young baboon. There has been much discussion as to why this particular lot of chimpanzees should indulge in such unusual behavior. Some anthropologists think that the explanation might be that they lead a rougher life than that of the rain forests, and don't get enough wild fruit to satisfy their needs.

Their use of tools is not as surprising—other chimpanzees have been seen sticking twigs into bee communities for the honey. Even there, however, Goodall's apes went further than the others, for they picked out leaves or bits of vine or twigs carefully and critically, and *prepared* their tools. When a chimpanzee had such a leaf, he would trim off the extra bits of it until he had left only the center, or if he had selected a twig, he carefully removed the leaves until he was left with a useful stick about twelve inches long. Having prepared the stick a chimpanzee might carry it a long way to the most promising terrain, such as a likely-looking termite nest. There he would thrust the stick down a hole, wait a little, and draw it out covered with termites, which he would nibble off much as a child nibbles ice cream off a Popsicle.

The Gombe Stream chimpanzees have also invented another kind of tool, to help them get water. Most wild chimps do not drink much water: here and there one has been seen sucking it up from a stream or pond. But these chimpanzees would take big leaves and crumple them, even chew them, then hold them in some stream until the leaves, like improvised sponges, were soaked, and the moisture could be sucked up.

Miss Goodall was still in the middle of her venture in 1962 when Vernon and Frances Reynolds began chimp watching in the Budongo Forest of Bunyoro in western Uganda. A map will show how close together the observers were, but the Budongo chimps (also *Pan t. schweinfurthii*) lived in what is a far more typical habitat than the Gombe Stream Reserve, being rain forest throughout. The Reynolds expedition took 270 days in all, from February to October. Their animals were more accustomed than were the Goodall chimpanzees to human company—even that of white people; a sawmill had been operating in their area for more than forty years, and several well-used man-made tracks ran through the forest. In the limited time they had, however, the observers failed to habituate the chimpanzees to their presence.

There were many apes in the district, and the Reynoldses rarely saw the same ones twice in as many days. Setting up blinds was no good; the chimpanzees kept away from such erections, and—as with Nissen—the group movements were unpredictable. The watchers therefore called on the services of friends who were willing to help keep tabs on the wanderings of different groups, and this worked pretty well. The Budongo chimps were primarily fruit-eating, but they also ate leaves, pith, and bark. Once in a great while they were seen eating insects as well, but they were never seen eating meat—or bird's eggs, either. (Dr. Kortlandt had tried leaving meat and hen's eggs in the path for his chimps to find. They always ignored the meat, and only one animal took the eggs back into the forest.) Nor did the Reynolds chimps use tools.

It rained a good deal during the 170 days, and the observers noticed that the chimpanzees' indifference to water turned to active dislike when it came down as rain. Nissen had never seen the Kindia apes in a downpour, but both Jane Goodall and the Reynoldses saw that their chimps hated rain. In a heavy downpour they hunched up in trees, "head resting on folded arms, head down" (Reynolds). Sometimes they vocalized in sad tones, as if lamenting.

It was in their observations of the group system that the Reynoldses made their most distinguished contribution to chimpanzee literature. They saw that in the Budongo Forest the usual bands were not closed or immutable; that the chimpanzees were "constantly changing membership, splitting apart, meeting others and joining them, congregating or dispersing." They watched one popular fruiting fig tree for eight days and kept a record of who visited it, in what company. Two mothers with juveniles came every day and often outstayed all the others. For five days a white-backed old male and a white-backed female came to feed—and so on.

The times of arrival and departure varied with the individual: never did all the animals arrive or depart as a group. Other chimpanzees were feeding at a place three quarters of a mile north along the river, and when the individuals left the fig tree, that is where they headed.

Once in a while the watchers recognized some small band that stayed together as a relatively stable unit, but on the whole the composition of groups was unstable. However, certain *types* of grouping were frequent, as Dr. Kortlandt noticed on the plantation. There the chimpanzees sorted themselves out in what he called nursery groups and sexual groups. The latter would visit the plantation only every two or three days, whereas the mothers with children were more faithful in their attendance. The sexual groups were more rowdy than the others, but between all these groups there was mixing and changing around, and on the plantation he saw little evidence of friction or resentment when such changes took place. Sometimes lone chimps would trickle in, he said—any sort: single males, single females, mothers with infants—and if they felt like joining up, that was all right, too.

But in the Budongo Forest the shifting and changing seemed to lead to a lot of noise, sometimes long after the customary bedtime hour. In the Ituri, Chimpo and I used to hear tremendous yelling and drumming at odd times of the day, way off in the trees—"Yow! Yow!" screamed the wild chimpanzees, with a great rolling sound like tomtoms—and we always paused in whatever we were doing to listen. Once, I remember, Chimpo had gone to bed when it began, and it was eerie to hear her answer with a small, gruff, tentative bark in the night. The Africans told me that the chimpanzees at these times were having a dance, stamping around in a circle (as the Africans themselves did) to the beat of a drum. Every once in a while, they said, one of the dancers got a step wrong, upon which the others would hit it and make it cry, hence the loud shrieks.

Back in 1900 Garner was told the same story. He heard that the noisy chimpanzees were having a carnival, drumming on a clay drum they had modeled for the purpose, and jumping up and down, and singing.

The Reynolds couple heard this "carnival" six times during their stay in the forest—four times by day and twice by night. In all six cases the noise was prolonged for hours. It seemed to be associated with the meeting at a common food source of bands that may have been unfamiliar with each other, and the noise might have been a part of their getting-acquainted routine. Once the watchers tried to locate the animals during the carnival, to see what kind of behavior accompanied such a tremendous uproar, but they found it impossible. Calls were coming from all directions at once, they reported, and all the groups concerned seemed to be moving about rapidly. The Americans heard stamping and fast-running feet sometimes behind them, sometimes in front, accompanied by outbursts of howling and the prolonged rolling of drums which shook the ground. After a while the most intense source of the racket moved off toward the south.

There seemed to be little struggle for dominance among the Reynolds chimpanzees. Sometimes there were signs of difference in status, but this occupied little time in their routine and caused less stress. In the words of the observers, there was no evidence of a linear hierarchy of dominance in males or females: no observations of exclusive rights to receptive females, and there were *no permanent leaders of groups*. This is so unlike the social structure of other primates, such as baboons or rhesus, that it is difficult to believe, but so matters appeared to the Reynoldses. They listed only twenty-five occasions when one chimpanzee showed dominance over another animal or a group. On seven of these, one male moved to allow another to pass or moved away as the dominant animal approached. Four times a small grooming group was seen to break up at the approach of a large male, and on one occasion a female who had been grooming two males moved up and groomed the intruder. Three times one male seemed to be the effective leader of a small band of adult males, for they left the tree when he left it, or followed him when he crossed the road.

There were always some adult males who showed "a relaxed and confident bearing and unhurried gait." One of this sort, a white or gray-backed male, might stay on in a tree after the rest had moved on, looking around, not at all worried if he happened to spy human observers—on the contrary, he sometimes showed an amiable curiosity. Where most of the adult males were very cautious before stepping into the road and then hurried across to the other side, these relaxed gray-backs (but not all gray-backs) had a way of sauntering across, not deigning to hurry.

Possibly, say the Reynoldses, the looseness and instability found in chimp groups might have been exaggerated. "In fact, one may entertain the hypothesis that chimpanzees possess a social organization so highly developed that it can persist in the absence of immediate visual confirmation normally true for baboons." When some of them find a new supply of food, for instance, and begin to make more noise than usual—might this not be a means of signalling to all other chimps within earshot, calling them to come and eat? And where vocalization doesn't carry far enough, drumming might.

One of the most dearly cherished projects of American primatologists, inside or outside government agencies, is to maintain somewhere easily accessible to themselves a colony of primates that live as they would in the wild. In a few instances men have come close to achieving this goal. One such effort is the colony of *Macaca mulatta,* or rhesus monkeys, that occupies the little "Monkey Island," Cayo Santiago, off the shore of Puerto Rico. It was founded in 1938 by the School of Tropical Medicine of the University of Puerto Rico in partnership with the Columbia University College of Physicians and Surgeons. Four hundred monkeys from Lucknow in India were turned loose at that time, and their descendants have been there ever since. Rhesus have quick tempers and possess a complicated, well-marked social organization that renders them interesting to animal behaviorists, though to my mind nothing can make them lovable. On Cayo Santiago they flourished, but they soon ate up all the natural food, and now they must be fed supplementarily. The colony has had its ups and downs, financially speaking, but at last, under the protection of the National Institute of Neurological Diseases and Stroke, it achieved stability. Now it is a mainstay of the Laboratory of Perinatal Physiology that is located in San Juan, and plans are under way to colonize some of the neighboring islands with other species.

The most ambitious of these projects for primate colonies is probably the one entertained by—of all things—a branch of the United States Air Force. An account written by a doctor and two Ph.D.s—*An Evaluation of the Behavior of the ARL Colony Chimpanzees*—carries the story a certain distance but does not finish it. The official name of the project was the Chimpanzee Colony of the 6571st Aeromedical Research Laboratory, Holloman Air Force Base, New Mexico 88330. The colony, near Alamogordo in the south of the state, has since been taken over by Albany Medical College. Holloman had a lot of chimpanzees, which until April 1966 served chiefly as a pool for space research subjects—Ham, the first chimpanzee in space, was a Holloman chimp. Then it occurred to some of the neurophysicists connected with the base that it might be possible to study what they called "the telemetered electrophysiological correlates of the freely moving primate or to study the behavioral effects of telemetered impulses delivered to specific neural centers and tracts through implanted electrodes."

However, they added, such studies would have to be done on animals with carefully documented behavioral baselines if they were to mean anything, and it could not be claimed that the chimpanzees at Holloman could be so described. Like all captive chimps they lived in conditions where they had no opportunity for social grouping and interaction. Typically they lived singly or in pairs, in cells much like those of human prisons. The reason for this treatment, of course, is that chimpanzees are very strong and clever, and it has always been easier to keep them segregated in this fashion, but for the purposes of studying behavior the method was most undesirable.

"Not only are the adults kept in isolation or semiisolation," wrote the authors of *An Evaluation . . .*, "but the young pass through critical periods of growth and development under the same impoverished conditions." Babies are often separated from their mothers early in life, if not at birth, and such infants develop "stereotyped behavior," rocking back and forth and hugging themselves closely, their hands clutching their own hairy skin for want of their mothers'. The authorities reasoned that it might be worth the trouble to create a chimpanzee haven at Holloman where the apes—of which there were many—could live together and develop social behavior in groups, as their brothers do in the wild.

9

A Consortium of Chimps

When an armed services committee goes to work on a new project, its first act is to dip into the alphabet. The 6571st Aeromedical Laboratory, having decided to confer on its chimpanzees an "enriched environment" (more living space), christened the addition the DPS, for Development of a Primate Source, Chimpanzee Consortium. For "Consortium" read "yard." A thirty-acre piece of New Mexican desert was enclosed by a circular water-filled moat, the planners reasoning that since chimpanzees in the wild avoid crossing streams or muddy ground, a wet moat would be a better barrier than would ordinary fences. Chimpanzees, of course, are very good at climbing most fences, but one such was put up at the consortium nevertheless, and furnished with an electrical charge like those used to control cattle.

Attached to the moated area, or island, was a shelter building of two feeding rooms, each with food hoppers, water spigots, and a wide shelf across one wall where the animals could rest. Out of doors the planners put a number of transplanted fruit trees and cottonwoods. Otherwise the only vegetation in sight was "the sparse low, shrub-type chaparral typical of the New Mexican desert." That was the consortium *in toto,* except for ten cement shade shelters.

The planners were hopeful that chimpanzees living in the consortium would exhibit social behavior like that of their species living free in the African forest. They and the apes together had to learn the hard way.

There was little similarity between the lush dark woods of, say, the Congo and the hard bright heat of southern New Mexico. Wild chimps come out into the open only when they must do so in quest of food. They are not thick on the ground, either: in the Budongo Forest the Reynoldses calculated only ten apes per square mile, and the Goodall chimps were one third as many as that. At Holloman Air Force Base, anything from twenty-five to forty-five animals were expected to live contentedly on thirty acres—the number varied because the authorities, wishing to reproduce the shifting and changing of groups that had been observed in Africa, occasionally removed some animals and introduced others. They were aware, of course, that the consortium was comparatively crowded, but reasoned that this would not matter since the chimps were fed twice a day and didn't have to live off the land.

Unfortunately the chimps did not reason in the same way. Food or no food, their nature is to tear up trees, and in no time at all they had ruined every tree on the island. In their native woods this presumably would not have mattered, since a few chimps cannot ruin a whole forest no matter how hard they work at it. As it was, within a short time nothing was left of the transplantings but a few dead cottonwood trunks, standing like Dali figures on the desert.

Moreover, the fact that the chimps were artificially provided with food—and were even fed twice a day instead of once—took a lot of seasoning out of their lives. In his natural state a chimpanzee spends all day scratching for a living: it is all he knows how to do, and everything else is subordinate to the search for food, even though he may lead a highly social life between meals. The consortium primates, deprived of most of this recreation, had to pack all their fun into those few minutes twice daily when the hoppers were filled with food and speedily emptied by the apes.

"Food time was a time of competition, great excitement, and a high incidence of a wide variety of behavior," wrote the authors of the report. "The rest of the time was really 'time on their hands.' Although the animals were free to roam the island at will they tended to remain in the feeding rooms and in the shadow of the shelter building to avoid the hot desert sun. Excursions away from the building were usually made in early morning and late afternoon and on cool cloudy days." Which was only, as a matter of fact, to be expected, for African chimpanzees avoid the direct rays of the sun. During their infrequent outings into full daylight, they sometimes wear large leaves on their heads, like sun hats. Even with these disadvantages, however, the authorities figured that thirty acres of land, however desert, would be preferable for the apes to a confined existence in separate cages. At least, the chimpanzees would have a chance to interact, they decided. And interact the apes did, often in unexpected ways.

Knowing how important background and conditioning can be, the scientists of the ARL colony tried to amass information about their subjects, but not a good deal was known about many of the animals because at the time they were acquired nobody had bothered to keep records. It was known that the biggest specimens had been obtained from zoos and circuses, where full-grown apes become a problem and the owners are glad to get rid of them. Otherwise, most of the collection, it is probable, were Africa-born and had been caught as infants in the wild: almost certainly they had undergone painful experiences at that time and later when they were sent abroad. It was bad enough for a baby to be snatched from his mother—who was very likely dead by that time—but what happened to him later was probably worse. At an animal dealer's in Nigeria, at Kano in the north, I once saw a small chimpanzee packed and ready for shipment by air in a wooden cube-shaped case so small that the creature had to keep its head tucked down almost to its knees, and when I asked when it was to be sent off, the dealer answered carelessly, "Oh, probably within ten days or so." Even if this chimpanzee survived, which seems unlikely, it would hardly retain any vivid memory of group life among its own kind.

Once they had ruined all their trees, the consortium chimpanzees' only playthings were bits of stick and branches from the chaparral and a few rocks, plus whatever implements they managed to steal from their jailers. In the rueful words of the authors, "Although the Consortium provided an enriched environment, it fell short of replicating their native African habitat."

For fourteen months they were observed thoroughly, all day, at periods ranging from a few minutes to three hours, until there was little the behaviorists didn't know about them. What with graduate students lending a hand in the summer and everybody making notes and taking pictures, the chimpanzees were fully aware that they were the cynosure of all eyes as long as daylight lasted. Even at feeding time when they appeared to be alone—for the windows of the building were of one-way glass—they must have known that a number of Big Brothers were watching them. As far as anyone could tell they were not inhibited by this knowledge. No one who knows anything about prison life will be surprised to hear that the consortium chimpanzees were a very quarrelsome lot, but the psychologists who directed the experiment, being scientists, did not come to this conclusion in advance, for that would have been theorizing on the basis of human behavior, which is wrong.

They remained firmly nonanthropomorphic, receiving as fresh the attested news that a "high incidence of aggression" was the rule, and commenting, "This contrasted sharply with observations of chimpanzees in the wild. Goodall reported that attack was seldom seen and usually resulted from disputes over food or acquired objects. Usually, attack was avoided by gestures and vocalizations. Scars and other evidence of past aggression were rare in the wild animals." In contrast is the consortium record. In one week there the watchers counted 108 aggressive incidents, most occurring around feeding time. Females were attacked more often than males, but the females too were aggressive, attacking males *and* females regardless.

Every time the group was altered, by withdrawal or addition, the chimps got as upset as they did when rain, wind, or dust storms suddenly descended on Alamogordo, and they became equally excited every time their dominant male came home to the shelter after a walk on the island. The fact that there *was* a dominant male marks another big difference between the social structure of wild chimpanzees and those of the consortium. At Holloman there was always a leader, boss, or dictator—whatever term one wishes to use—and he had to keep alert so as not to lose his position. Often during the frequent bouts of excitement, the watchers saw aggressive acts between two animals turn into mob war, with whole groups of chimps attacking both the first aggressor and his victim. Newly arriving animals were usually immediately attacked. Sometimes they were only mildly harassed, but on certain occasions the newcomer was so severely beaten that human attendants had to remove him. However, if such a one made the appropriate "submissive gesture"—offering the back of his hand or presenting his rear—he usually avoided being attacked at all, and was accepted into the group.

Animals that screamed or whined constantly or threw temper tantrums were frequently beaten up. (A chimpanzee temper tantrum is curious. The animal screams, stamps, rolls on the ground and sometimes rushes around blindly, bumping into whatever stands in his way. In anthropomorphic language, the chimpanzee is hysterical.) The annals of the consortium are full of haunting stories about individual animals like George, a large, mature chimp who tended to be neurotic. The behaviorists moved George out of his cage and put him into the consortium, which at the time was under the thumb of an animal named Sampson. During the first day a showdown was averted because George simply huddled under the turnstile where he had been injected into the colony and wasn't perceived. Next day, however, he was shooed out into the open on the island, and was promptly beaten—not by Sampson himself, but the others. Sampson simply "circled the melee" and "seemed to be directing the action." Poor George ricocheted from one attacker to another until, trying to dodge them, he slipped and fell into the moat. Chimpanzees are top-heavy, and George would have drowned if the observers had not pulled him out. Back on dry land, he meekly let himself be led by the hand to his quiet, safe cell.

At another time when the consortium leader was an animal named Chuck, Tim was put in. Like George and Chuck and Sampson, Tim was a fine figure of a chimp. Chuck and Tim met and began circling each other, both screaming continuously and making "submissive faces," grinning widely so that their gums showed. They never came to grips with each other, but now and then one of them would hug a nearby onlooker among the chimpanzees—a gesture often made by fearsome or worried animals—until at last Chuck presented his hind quarters and Tim mounted him for a minute. (Mounting between male chimpanzees is a sign of dominance, not sex.) Chuck then turned around and even more specifically invited Tim to be the boss, presenting the back of his hand, limply held, to Tim's mouth—the chimpanzee equivalent of saying "Uncle." Tim took the hand gently in his mouth and the two animals walked along together for about ten feet before he dropped it. Power had passed from Chuck to Tim, and the watching chimps, who all along were shrieking as if in terror, now calmed down and were quiet.

In nature chimpanzees are not what we consider sexy. When they do copulate they are casual and promiscuous, never showing sexual jealousy: Such incidents as have been reported are more often than not the result of female, rather than male, enterprise. But the apes of the consortium were different: they "exhibited an extremely high frequency of copulatory and other sexual behavior," males as well as females taking the initiative.

Three healthy chimp babies were born at Holloman in 1966. One of them, Susu, was kept with her mother through the first year of her life. She was a well-adjusted little chimp, sucking her thumb only once, and that briefly, soon after birth. "She formed no intermediate object-relationships and was first observed to show a tentative interest in potential play-objects at 9 months of age," wrote the authors, by which I think they mean that she didn't hang on to a security blanket and began to play with toys at nine months. Phyllis and Little K, the other infants, did not fare as well. Phyllis, who was separated from her mother at birth, immediately formed an intense attachment to a blanket: at nine months she was still carrying it everywhere. Whenever she was upset, especially when the blanket was taken away for a time, she sucked her thumb. Little K, who was separated from his mother at five months, also fell in love with a blanket. He wasn't as firmly attached to it as Phyllis was, but he too wouldn't go to sleep unless he had it. Little K sucked his thumb less often than Phyllis.

One of the difficulties encountered in keeping chimpanzees is their love of fiddling with things, a practice regrettably described by behaviorists as "object manipulation." At Holloman when caged animals had nothing better to play with, they arranged and rearranged bits of food in nests around themselves as they ate. When this palled, they wiggled the padlocks on the outside of their barred gates. Even when they had discovered that the padlocks wouldn't break, they rattled them. The comparatively free animals on the island brought back to the shelter any playthings they could find—burlap bags, metal rods, and building tools left by workmen. They also threw rocks and sticks at each other, their aim improving as time passed. About half the apes learned from the other half how to spit, and so time was whiled away through one stratagem or another until somebody thought of playing in the moat. Until then, the island's inhabitants had behaved toward the moat much as their guardians had expected them to, remembering that chimps are thought to have a fear of water. But one day a flung stone missed its mark and landed with a splash in the water: chimpanzees promptly threw in more objects for bigger splashes, and after a while a brave ape dipped a stick into the moat, pulled it out, and licked the water off. Others followed his example, and one thing led to another until in 1967 a daring young female learned that she could make splashes by hitting the water's surface with her hand. Soon she was pulling things out of the moat. Then she found out that she could lower herself *into* the water quite safely as long as she held on to the cyclone fencing that covered part of the side of the ditch. Dipping then became a regular form of recreation at the consortium—but its originator, the daring young female, drowned.

That was as far as the story went in print. In 1968 I flew out to Alamogordo to see the consortium for myself. Getting there is a fairly complicated business entailing changes of plane, because the air field is too small to take ordinary national carriers, but I got there at last, landing on a sandy plain that lay prone for miles between two distant blue mountain ranges. In the slanting rays of an autumn sun the sand sparkled. The few buildings that stood near the airport were dwarfed by the sky's enormous stretch. An officer, Captain—or Doctor—Robert McRitchie drove out to fetch me. (As all the officers I met at Holloman were medical men, I shall ignore their military rank.)

He said as the car started, "I hope you aren't expecting great things at the consortium. There's not a lot to see now, because the original idea kept running into snags and we also had a considerable amount of bad luck. The moat, for instance—Who would have expected chimps to get used to water? But they did; they even began wading right across and then taking off as fast as they could go. We had to deepen the ditch. Even so, one animal seemed determined to commit suicide. And sometimes they tried to drown each other: we lost a few like that. I tell you, it was a mess. You can see the moat over to the right if you look fast."

I did see it, curving away in the distance, just before our road curved in its turn and took us into and through a number of gates and a complex array, like a little hamlet, of buildings. "There isn't any water in the moat now, is there?" I asked.

Dr. McRitchie shook his head, a little grimly. "There isn't; that's right. We had trouble with the lining of the ditch—it cracked and the water seeped out. One way and another it's been some time since we kept chimps on the island. There was another time during the winter when we had a freak cold snap and everything froze over: all they had to do was run across without even getting their feet wet, and some did. The island's been a headache. Thank God, it isn't my department: my principal job is coordinating immunological material for other places, chiefly New York. We work in with Dr. Alexander S. Wiener and Dr. Jan Moor-Jankowski.

"You know, don't you, that chimpanzee blood is in some respects like ours? So is baboon blood. We send sera all over the world for comparative studies. One of our programs is to establish a kind of serum bank of all the known blood types of chimpanzees. The plasma can be frozen and preserved in liquid nitrogen. There's a prejudice against using primate blood for these purposes, but if we could overcome it, we have a most important potential."

He slowed down the car as we passed within sight of a large baboon in a cage just inside a shed. "That animal must just have arrived. He's a beauty, isn't he? I wonder who's working with him. . . . Look, over there is the original Stapp speed bench." It looked like a pair of giant's skis with a small chair at one end. "Do you remember the tests they used to make with that machine?" he asked. "The effects of impact, and so on—why we must use safety belts, and what happens in space."

"I think I remember," I said. "There were photographs of chimps in space helmets, buckled into chairs like that one."

"That was it."

We parked the car and entered the main building. Here there was no immediate evidence of living chimpanzees, though Dr. McRitchie assured me that there were at that date 138 of the animals nearby. All was brightly lit and businesslike, from the headquarters of the officer in charge to the little auditorium complete with screen and projector, where I saw movies of space experiments and learned something of the work being done by the ARL at that time, in the dawn of the Space Age. If here and there I missed a point because of the unfamiliar vocabulary used by these medical men, it was not their fault: I should have asked—and sometimes I did. I looked respectfully at a list thrown on the screen for my benefit:

PRIMARY AREAS OF INTEREST

Primate baseline studies
Calcium-calcitonin effects
Bacteriology-virology primates
Blood groups
Immunological factors
Comparative neurophysiology
Work-rest cycles
Sleep research
Cardiopulmonary studies
Experimental embolism
Implanted brain stimulators
Physiological effects—impact
Tolerance—impact, deceleration

At this point I stopped for breath, surprised that there was nothing really impossibly mysterious about it all so far. I plunged on:

Develop performance tests
Develop performance equipment
Comparative psychology
Comparative visual processes
Comparative circadian rhythm
Comparative pattern recognition
Decompression
  physiological effects
  performance decrement
Time of useful consciousness . . .

Here I broke off to ask, "What does 'circadian' mean?"

"Oh, that's your everyday ordinary routine—*circa diem.* When you fly a long distance, or rather when you're in the air a long time—across the Atlantic, for instance—your sleep pattern and digestion become upset.

The routine's been interrupted. A lot of these effects and others aren't thoroughly understood yet, but we know the upset takes place."

"I know," I said. "It often takes me as much as four days to get over that funny feeling I have after a transatlantic flight."

He replied cheerfully, "It's more like two weeks before you're back to normal, but you're not aware of that."

By a natural transition I went next to meet an enthusiastic young man in charge of sleep research. "We've had a sleep laboratory here at Holloman for about three years," he told me, "which means we're comparatively new as laboratories go. We began with rhesus because their sleep stages are something like our own. There are five stages of sleep, you know. First stage is a state of drowsiness, when you're just dropping off: with stages two, three, and four you get progressively deeper into sleep. Stage five is what we call dream sleep, at which a change occurs. It's not necessarily deeper than stage four, but it's different. The whole cycle lasts from sixty to a hundred minutes: a person sleeping normally will run through several of these cycles each night, possibly five or six, all the way from light sleep to deep to dream. You've probably heard that a dream lasts only a split second, haven't you?—no matter how long it seems to take. Well, it's false: a dream can go on for quite a long time. The reason people got that idea about split-second timing is that they so often forget their dreams after waking. If a dreaming person is aroused in the middle of a dream he doesn't forget: he remembers it clearly, but if he's had his dream out and finishes it, he may not recall any of it next day. One aim in our research is to discover how to achieve stage four in a shorter space of time. Stage four—deep sleep without dreams—seems to be the recuperative, restorative stage. A sleep-deprived rhesus or chimp, kept awake for twenty-four hours and then permitted to sleep, goes into stage four almost immediately.

"There's no doubt about it; we need to find out a lot more about the nature of sleep. Why do we need sleep? Are there specific areas of the brain that trigger it? Then there are the effects of chemistry, the different drugs that bring on sleep or wakefulness. In these matters we can control monkeys better than humans: they can't trick us as the humans can. We'd like to find out if they recover from deprivation in the same way as humans. The record time a rhesus has been kept awake is eight days compared with the human record of eleven, and we didn't notice any neurological damage in either the human or the monkey. There's no doubt that with us a good deal of fatigue can be overcome with high motivation, but you can't sustain the performance forever. Yet the animals that went eight days without sleep worked at their ordinary tasks for food in the usual way, though they looked just about dead, and certain animals *have* been kept awake until they die, I believe.

"I was struck by one feature of the last space flight. [This was Apollo 7.] Out of four things the astronauts complained of, three had to do with sleep. They said their sleep was interrupted too often, that they didn't get enough sleep, and that their sleeping arrangements were uncomfortable—the last squawk was about food. So there you are, three out of four. Why, since we're getting to know so much about sleep, do we send them up to suffer as they do? The sleep hardware in the space module is really awful—the net, the belt that holds them in, their hands being tied down—of course, they can't sleep. They ought to have more comfort. Perhaps one of these days we'll work out a method of control by which men in such situations can get enough of the right sort of sleep."

Our last call was on Dr. Donald S. Farrer of the psychology department, who told me it was from the ARL ape collection that the Gardners of Reno obtained their famous chimpanzee Washoe, now learning to use a sign language.

"This laboratory is purely for applied work," he said. "We give the animals problems, using the Skinnerian approach [i.e., rewards for good behavior, but no punishments], and see if and how they can be worked out. NASA asks for certain answers, and we're here to supply them if we can. But one of my own favorite projects deals with the theoretical-cognitive behavior of the chimpanzee."

At my expression of frantic noncognition he relented and slowed down. "For example. We did a nice little experiment a few years ago when we had to develop a fairly simple task for the animal. You may recall the experiment used by Mrs. Kots in Moscow in 1910 on her chimpanzee—the matching-sample test to discover whether a chimpanzee is able to distinguish colors. We elaborated on that and developed an automatic projector with which we show the animal a sample of something, in this case a shape sample, not a color one—a triangle, say. The chimpanzee is seated before a board on which various shapes, including a triangle, are painted on as many buttons. If, on seeing a triangle projected before him, he presses the matching button with a triangle on it, he gets a reward.

"This started out as a comparison test, but what it turned into was a test for length of memory. In the full-dress study we have twenty-four different problems to solve. One of our chimps was very good indeed at it, so good that she became a sort of show all on her own—we always trotted her out when we had visitors because she nearly always pushed the right button. Then we began to wonder about her when one of us noticed that she always watched the visitors before she punched the button. Did her response depend on the visitors' reactions? Or—another possibility—had she memorized the whole sequence of symbols? If so, the test was of her memory rather than her powers of recognition. We tried her out without visitors and with a scrambled sequence of symbols. It made no difference: she got the right answers just the same. We decided that she must be an extraordinary animal, a genius among chimps, and to prove it we tried out two other animals, training them the same way. Do you know what happened? Those two new chimps got the right answers just as often as the first one did. They really could memorize those pictures: we had to admit it.

"Next we tried them all on partly obscured symbols, and they still did fairly well. In fact, we didn't flummox them until finally we turned the pictures around and presented the chimps with a mirror image of the shapes they were used to. At last, this did it: they acted as if they'd never seen the pictures before. They had to go back to the beginning and learn them all over again.

"We're often asked how clever the chimpanzee is. To my mind the proper answer is, How clever are *you* at asking the chimpanzee questions? It's establishing communication that often takes most of the time: sometimes you need six months just to make the animal understand what you want him to do. Then there's the matter of temperaments, your own as well as the chimp's. One investigator may find it impossible to work with a certain animal we'll call Martha: he'll want us to get rid of Martha because, he says, she's hopelessly dumb. Then along comes another investigator who finds Martha absolutely brilliant: he always asks for her. You can never tell in advance. It's like teachers in a school: no teacher treats all his students alike—it's really a question of rapport. Experienced workers here will tell you they can see from the beginning how an animal is going to react in a given situation or problem. A very active, aggressive rhesus may learn to press a lever more quickly than a docile animal would, and prove more easy to teach.

"Incidentally, a rhesus can undo a latch at first sight of it. Here's an interesting question, Can you define manipulation as a basic need? It's an almost universal aspect of primate behavior. *Man cannot create any machine that primates cannot take apart.*" Dr. Farrer grinned. "That idea's pretty controversial, I guess, but it's fun to play with. Another question we're often asked is whether or not chimps can count. My own belief is that they can, at least up to five. A counting chimp can be trained to push a button marked five, as a kind of stopping place, or punctuation. Some people object to this as proof, saying that the chimp simply recognizes the symbol for five, and we rigged up a new set of symbols instead of our numerals: the chimpanzee still got it right. Admittedly you'd have to define the term 'counting.' To avoid this difficulty we might say the animal has a differential response to five different symbols. . . .

"Our animals play a cage-to-cage version of tic-tac-toe with nine symbols and an electronic board in each cage, outfitted with a sign that tells each chimpanzee when it's his turn to play. Sometimes a player cheats by pushing a button twice, instead of once when it's his turn. And sometimes a chimp becomes disgusted and gives a Bronx cheer. The one that wins gets a food pellet or something like that. One of our animals got very good, winning every time: another always blocked his opponent. In fact, that second one, Paleface, became so good that we had to discontinue the games for him when he attained the weight of ninety-eight pounds and learned to push his Plexiglas window out in the excitement. The experiment wasn't worth the risk and expense. Another chimp, Big Mean, once got so mad after losing three or four games to Paleface that she wouldn't play at all, but sulked and refused to push the button. This spoiled Paleface's game until he learned to let her win a few times."

Dr. McRitchie took me on a quick visit to the caged-animal department, depressing as an outsize monkey house in any old-fashioned zoo, with one strong cage after another. Some held animals of impressive size. The cages were large and the place very clean, but it was gray and stank of antiseptic, and soon we went out of doors to look at the deserted island.

"You could call it a noble experiment, I guess, while it lasted," said my companion, "but sometimes there was hell to pay in there. You see that fencing over there? The idea was to cut the area in two so we could observe two lots of animals. They thought it would work fine—it's electrified, or was, and as you can see, it's made of reinforced piping with cables that carried the charge. It didn't work for as much as five minutes. The first chimp to get around it did just that—he grabbed the pipe at the end of the fence where it overhangs the moat, and swung around to the other side, evidently without getting the slightest shock. The others climbed over wherever they felt like it. They have thick skin and hair on top of that, which seems to insulate them. One female climbed the fence after she'd run into the moat, while she was still soaking wet, and the current didn't affect her at all. Before they built this fence they had a weaker one—the chimps pulled it up and used pieces of it to make nests. It's surprising how much strength they have. Now we're designing an even stronger fence; it just might work."

We came to a place where a sort of avenue had been constructed with a double line of fencing across the middle of the area. The pathway thus left was about seven or eight feet across. Dr. McRitchie asked the guard who accompanied us, "Who's supposed to walk down there when you've got animals on either side? I wouldn't, I can tell you."

The guard grinned and shook his head. "I wouldn't either," he said. "They're trying something better now."

Since my visit there has been a new publication by the ARL people: *Colony Management and Proposed Alterations in Light of Existing Conditions at the Chimpanzee Consortium.* The authors make several points: that it is vitally necessary to insure a future supply of chimpanzees, considering their decimation in the wild; that to insure the consortium apes' health, fighting amongst them should be reduced; that "there should be an increased effort to consider the manipulative capabilities and strength of chimpanzees when planning future construction," and that better use should be made of "the research potential of the facility." The authors say that there has been "insufficient communication between personnel with knowledge of these capabilities and those who design and construct the structures," and they add, "If a chimpanzee can reach a bolt or loose lock, he will remove it. If there is a 'cold' spot on a high voltage structure, the chimpanzee will quickly discover it; their thick skin and hair has high resistance and provides an insulation against shock. Furthermore, chimpanzees have a propensity to build nests . . . and will use all available material to do so. Their great strength even permits them to use cyclone fencing if it is not firmly secured. During periods of great excitement chimpanzees tend to hit and slap nearby objects including observation windows within reach. Because of a scarcity of shade and manipulable objects on the land area destruction has been most intense in and near the building. The moat has been successful in preventing escape, but several animals have drowned there."

In addition, they remark, because food, water, and most of the shade from the desert heat are available only at the shelter, the animals naturally congregate there and don't use the rest of the area. Also, in comparison with wild groups there are not enough adults to deal with the adolescents at the consortium. If two chimps don't like each other, they can't get out of each other's sight there as they could in the wild.

To meet some of these objections the writers suggest that feeding and watering stations be placed all over the area, so that newly introduced animals would be able to get their rations without facing the whole community, and there would be fairer shares for all. The land area should be divided into halves by a moat rather than the present fences, and both sides should be provided with "sturdy concrete and steel playground equipment," possibly including jungle gyms and cement sewer pipes.

There should be much more shade to relieve the congestion around the shelter, and introduction procedure should be changed: instead of simply throwing a newcomer in, he might be introduced by slow stages, while the dominant male is somewhere else. Not that all the "status" males turn out to be bullies. Tim, for instance, was more of a policeman than a tyrant, generally breaking up rather than initiating fights.

10

An Enriched Environment

Anyone who has ever tried to raise a chimpanzee can bear witness that it is not easy. In the words of the director of the Baltimore zoo, who kept a young one in his house, "I've had him here for two years, and that's a lot of chimp."

So I paid attention when I heard for the first time of Professor William B. Lemmon, of the psychology department of the University of Oklahoma at Norman. "You'll be interested," my informant assured me. "He keeps chimpanzees. I think he observes their behavior."

"Is that so?" I asked politely. "As a matter of fact I saw a couple of chimpanzees only yesterday in Washington—young ones, of course."

He said, "Well, I don't know how old Bill Lemmon's are, but he must have all ages, big and small. He's got some whoppers."

This was what made me prick up my ears. I asked how many there were at Norman, and jumped at the reply: "Probably about twenty, I guess. Why, is that unusual?"

I got busy right away on what I could find of Dr. Lemmon's writings, and came up with a paper entitled "Delivery and Maternal Behavior in Captive-Reared Primiparous Chimpanzees, *Pan troglodytes*." He had been working, it seemed, on the familiar question of the nature of what we used to call instinct—or, as he put it, which aspects of complex social behavior can be considered "pre-wired" and/or learned. "Maternal behavior would seem an excellent area for such investigation," he wrote.

"Its species survival importance is obvious and maternal behavior is clearly composed of molecular components whose molar integration necessarily varies with circumstance."

This was difficult stuff, but I looked up his use of the word *molar*—"mass, opposite to *molecular*"—and doggedly went on reading. The study began some time ago, he said, when he and his associates first acquired a pair of wild-caught chimpanzees, each about a year old. During the first year the apes spent with their new owners, they demonstrated many "integrated, socially significant behaviors which seemed to occur without demonstration or practice." But the humans felt that one could not be sure. The chimpanzees might have observed and learned this behavior before they were captured, "the particular behavior remaining, as it were, quiescent until an opportunity arose for its direct repetition." The human owners decided therefore to obtain a few newborn female infant chimps, and "to investigate sexual and maternal behavior under circumstances where direct learning or initiation could be controlled."

This sounded good, but as they discovered, it was not at all easy to get newborn chimpanzees, male or female, and it was only after a two-year search that they acquired two such animals the day they were born. The chimpanzees were half sisters. "Each was born, six weeks apart, of an ex-performing mother and a common sire," records Dr. Lemmon; and each was immediately placed by adoption, as it were, with "a faculty family," in absolute species isolation—or as we laymen would say, in situations where there was no chance of either infant seeing other chimpanzees. Since then, said Dr. Lemmon, for two years their social behavior and development had been closely followed and observed.

But it seemed clear that if he wanted enough chimpanzees for future experiment, he would have to breed his own. He set to work, and within another four years he had collected thirteen more animals from zoos and dealers, as well as animal trainers and families that could no longer control their chimp pets. Each animal, as it happened, had been born wild—after all, chimps have been breeding in captivity at a reasonable rate for only a short time—and several of the animals had had no contact with their own species since capture. Some of the males, possibly as a result of such erratic development, were sexually inactive, if not impotent, but in the end this did not prove to be much of a problem to the well-being of the colony as a whole. Within the year, three females had successfully given birth to healthy infants. Soon after the birth, each of the mothers gave her baby a mouth-to-mouth treatment, as in the "breath of life" technique, evidently to establish breathing. Other observers have seen similar behavior at various places and times, but the fact was surprising to the Norman people, nevertheless, as were other details of behavior. For example, none of the mothers put her infant spontaneously to the breast: the infant had to find it himself. "The mother is initially tolerant of nursing, and never becomes more than passively facilitating," reported Dr. Lemmon.

A little chimp's cry of distress ("distress vocalization") is an "imperative signal" to the mother, even if she doesn't know at first what to do about it. Twice the observers took a baby from its mother, then later reunited them, and the mother in both cases didn't seem to recognize her baby until it cried, or vocalized: then all was well. But there was no evidence, the professor said, of "felt loss" on a mother's part if her infant was removed from her. She didn't look for it, and there was no apparent grief felt. Mothers whose infants had been removed were not interested when they were offered stuffed animals instead; the other adult chimpanzees seemed scared to death of these inanimate objects.

An infant left with its mother responds early to chimpanzee vocalization (noises). From the first day it responds to food hoots with food hoots, to greeting hoots with greeting hoots even when the latter originate from an animal not its mother.

When the infant is about five months old there is evidence of separation, frequently initiated by the mother—which means, I suppose, that she encourages the baby to let go of her once in a while, and walk by itself. By the age of eight months the baby has learned to beg from its mother "with rather more arrogance than do adult chimpanzees." Little chimps take liberties with all the adults, but the mature apes are invariably gentle with them, even those not their own, and even when the babies are making pests of themselves. "Adult females, particularly, behave in a gentle, reproving fashion under these circumstances—in a fashion which cannot help but be socializing for the infant," wrote Dr. Lemmon.

Opposed to this pleasant fact is the disturbing one that the maternal behavior of the animals disintegrates under extreme stress. If a mother and infant are put into a squeeze cage—and if the mother has already been in one of those frightening contraptions, in which one of the sides can be moved, squeezing the inmate into immobility—in her agitation she usually drops the baby and ignores its cries of distress.

Summing up, Dr. Lemmon said that if we accept the observation made by Dr. C. H. Kratochvil, formerly of Holloman Air Force Base, that "below the level of the cortex, the chimpanzee's central nervous system is impossible to distinguish from that of the human by any physiological measure," it would seem possible to raise the question as to whether there are not many more "pre-wired," socially significant behaviors in both Homo and *Pan* than it is now fashionable to assume. "When we have had an opportunity to record the maternal behavior of chimpanzee females who have had no chimpanzee contact since the day of their birth," he writes, "we can certainly be more definitive."

My meeting with Professor Lemmon took place in his office in Norman. He is a large man with informal manners, and he wears a genuine under-the-chin goatee, evidently cultivated so that he can pull at it while talking or lecturing. He tugged at it now, sitting back comfortably in his chair.

"There's been a lot of crap circulated about primate behavior. The present god of behaviorists is Skinner, as you must have discovered by this time. He's got a really bright gang working there. Yerkes started it all, of course—he was pure, quite pure, and he had all the best people of his time. I'd say that at least half the psychologists worth anything today were once at Orange Park. After all, Yerkes had the only chimpanzee reservoir in the world after Köhler's laboratory packed up. He trained his people to be objective, *purely* objective"—I suddenly realized that Dr. Lemmon's eyes were fixed on me coldly and defiantly; he swept on—"and as a result they *missed a lot.*"

"I'm sure they did," I said heartily. He relaxed.

"He had this terror of anthropomorphism. All his people came out of Orange Park with a clinical objection to applying the principles of primate behavior to human behavior. But you've *got* to."

I nodded with genuine enthusiasm, and he took a fresh breath.

"I used to work in federal prisons here for a while as a consultant. It was fascinating. I found out a lot about the Oedipus myth while I was on that job. Freud was right, you know, absolutely right after all—if for the wrong reasons. The Oedipus thing is there all right, in our origins. If I'd doubted it before, the prisoners would have taught me it's there, but I didn't know until I moved on to animals that it's biologically determined. It's wired in, the whole thing."

I said, "See if I've got this straight. I can hardly believe—you mean *chimps* have an Oedipus complex?"

He nodded seriously. "I'll give you a copy of something I've written about this, but we'd better talk about it first. Right now I think we'd better go and see the family—we keep them in the country, of course. Ready?"

It was a bright September day. We drove through rolling country, in which I could see very few houses; even fences were not much in evidence. After some miles we turned in at the Lemmon driveway, while he explained that this piece of country was good for his purposes: he needed plenty of ground, not only for various kinds of primates but for other species he worked with, and there was nobody close enough to be bothered by chimpanzee noises.

He stopped the car near a white farmhouse with a cluster of buildings behind it and a big cage farther off to one side. On our right was a wide ravine, or arroyo, that might at times be a stream bed. Its red Oklahoma soil supported bushes, tall grass, and on the other side of the ravine a pretty copse of trees, like a little park, overlooking a pond. The trees were surrounded by a fence, behind which an animal suddenly leaped from one branch to another. It could not be a squirrel, I said to myself —the branches shook too much for that. I squinted for a better look.

"You're right," said Dr. Lemmon. "They're gibbons: I have a pair, and she's expecting an infant one of these days. We can take a closer look later."

He led the way around the house to the outbuildings, pausing on the way to greet a young man in a white coat. Another youth in blue overalls opened a door for us, and we stepped through into a large, shadowy room and, at the same time, into a many-layered noise so palpable that for a minute I had the illusion that I couldn't even see—a deafening uproar of screams, roars, and the bong, bong, bong of a huge iron drum.

Directly in front of us was a skeleton wall of strong iron bars running from floor to ceiling. The bars formed a room within the room, and inside that barred room were smaller rooms, all walled off by bars. Here and there was an opening in one of the sections, like a door; this door could be shut by a large flat piece of iron. It was on such an iron that the chimpanzees hammered at times with their hands and feet. But these details I made out later. At first all I saw were the faces peering out at us, between great fists grasping the bars—chimps, a large number of chimps in all directions, staring raptly at Dr. Lemmon, though now and then a caramel-colored pair of eyes slid over to look at me for a flickering moment. Some of the animals stood up or at least stooped in a standing position. Some chose to leap up and hang high. In one corner a mother hung carelessly, one hand on a vertical bar and the other on the ceiling, her baby peering out from behind her shoulder. The baby, too, watched Dr. Lemmon, who kept talking, pointing the animals out and naming them to me, giving little biographical sketches.

"This one used to work in a circus, and he still does tricks sometimes for his own amusement—or maybe it's for mine. Here you are," he said to the chimp, and handed into his eager black fingers a lighted cigarette. The ape took it delicately and put it into his mouth, lit end first.

I cried, "Oh, don't! He's hurting himself, he'll burn himself!" But Dr. Lemmon signalled me to shut up. A moment later the chimpanzee opened his mouth and showed us the cigarette in the proper position, with the fire outward, then began to smoke it, inhaling with manifest pleasure, his dark eyes surveying me with a touch of complacency. Another ape banged the metal door violently and several animals screamed as if in tune. In a rear compartment water was running out of a faucet near a younger animal that had just turned it on. Dr. Lemmon roared, "Turn that off!" and the young chimpanzee hastily obeyed.

I said, "But he understood you!"

"Of course, he did. They understand a lot," he replied. "Yerkes didn't give them enough credit—he was so afraid of the wrong interpretations—but I assure you that a chimp can put two and two together and get twenty-six if he feels like it. Mind you, they won't obey just anybody. They wouldn't obey you, for instance."

As he talked he moved farther into the room, mounting a couple of ironwork steps, like fire-escape steps, that led to a passageway. I followed, thinking it was expected of me, until a long, wiry finger with a heavy curved fingernail came up from beneath and hooked itself around the side of my shoe. For the first time I realized that a chimpanzee was beneath the floor, which was hollow, and I screamed and leaped just in time to retain my shoe. But as soon as I landed on another spot, *two* hairy fingers poked up and this time nearly succeeded in stripping my foot before Dr. Lemmon's shouting sent the hidden chimpanzee scuttling out into the open cage. There he sat down and looked pleased with himself. Another animal pounded on the metal, sending out great waves of percussion, and yet others shrieked. Before the clamor died down, we left. Two chimpanzees near the door had to be diverted with bananas so that they wouldn't spit as I passed.

Dr. Lemmon told me that some of the chimps who were small when they arrived had spent a few weeks or months in the house with his wife and himself, to get accustomed to civilized life. "One funny thing about that," he said, "was the way they remembered. I keep a lot of plants in the house, and they were taught not to meddle with them. Then the chimps got bigger and went out to join the others. The other day we had a jail break—one of the assistants was careless about shutting a door, and a lot of them got out. My wife and I were in town when it happened, and for a long time nobody could get in touch with us, but finally we heard a radio broadcast, and I can tell you, we skimmed the top of the hedges on the way back. Nine of the chimps had gone through the house, and everything was a mess—except my plants. They hadn't touched one of them, though outside they rooted up everything in the garden. We used to have an exercise yard, but they ruined it. We'll have to build another when we have the money."

I asked how he had got the big chimpanzees to go back, and he looked surprised and a little offended. "I took them by the hand and led them in, of course," he said.

The cage on the roof held a mother chimp and her young male baby, a lively little animal. "Oh, yes," I said. "That reminds me. What was that about the Oedipus myth?"

"I mean that none of the male infants I've ever watched show the slightest sexual interest in their mothers," he said. "They start very young investigating other females, but they never make advances to their own mothers. There seems to be an inhibiting factor, and it's something very deep. And it doesn't work the other way around—mothers sometimes behave seductively toward their sons, but the sons ignore those advances."

"And fathers with daughters are the same?"

"No, that's different, but you'd expect it to be. How is a chimpanzee to know if a young female is his child or not? It's the mother-son relationship that's so interesting. I maintain that the taboo is wired in. This doesn't mean that male infants don't feel protective about their mothers: they do. If you threaten a mother chimp, her son will fly at you. But the female infants don't behave like that."

He talked about stereotypies, the odd behavior shown by young chimpanzees that are deprived of maternal care—rocking, thumb-sucking, and crying—while we walked around the grounds and looked at his other primates. In one lot of cages lived a colony of stump-tailed macaques, who, Dr. Lemmon said, have a built-in masochism: they pluck themselves when they are left alone. "Sometimes they even hallucinate," he said. "Now over here is another species—the *nemestrina,* or pig-tail macaque. It's quite striking, by the way, how these animals differ in one respect. Physically and in most behaviors they aren't very different, but in this one way they are. The female of the *arateides,* or stump-tail, makes a very permissive mother. You can see for yourself that they let the children run all over the place, playing with the others from quite an early age, but pigtail females are the exact opposite: they're clutchy. I call them Portnoy mothers. Look at those two, hugging their babies to within an inch of their lives. I'm considering an experiment, to arrange for some of each species to have babies at the same time; then I'll exchange the babies and see what happens to their behavior as they grow up. They might adapt to the very uncharacteristic behavior of their foster mothers, but they might not."

William Lemmon moved into psychology as a side effect of his interest in medicine, and into the study of animal behavior, in turn, as an outgrowth of the study of psychology. His first independent work was with waterfowl, from which he transferred his attention to pigeons because these birds are monogamous, and he was interested in monogamy.

"It's a long way from the subject of primatology," he admitted. "The only monogamous primate is the gibbon, but then he's got a low sex drive anyway. He can be a very vicious animal," he added thoughtfully, pulling at his goatee. "If pigeons hatch out of foster eggs, they always pick out mates of their own color, did you know that? This might be Oedipal, but—it's a complicated question.

"After about five years I got tired of pigeons. My daughter, who was five at the time, was raising a cosset lamb, that is, a pet lamb, and I started to watch it; then I studied a lot of sheep. The lamb lived in the house and was bottle-fed, and thought it was human. It was so human, in fact, that it didn't exhibit any of the fears usual in sheep. I wrote off to the NAS for a grant, got it, and worked out a paper on the hand-rearing of sheep. I took lambs and kept them away from their mothers and tried out an experiment: I would put them, one after another, on a glass-topped table, one part of which was covered. Some of the lambs would step out and walk right over the uncovered part of the table, and would probably have gone on trying to walk on air if I hadn't stopped them. Others, though, were scared silly and wouldn't move a step. It all depended on what had happened during the first ten hours of the lamb's life. If he'd been reared by his mother for even that short amount of time, he was a smart animal; but if he hadn't seen his mother at all after birth, he was dumb. He had no depth perception. Another thing: a lamb kept in isolation simply dies. It loses all appetite, develops muscular atrophy, and that's it. Living creatures seem to go through certain critical periods—'There is a tide in the affairs of men,' and 'A time to live and a time to die,' and all the rest of it.

"Well, I thought I might as well extend my research to goats, which after all are very like sheep though they can't interbreed, so I got some angoras and tried out the glass-table experiment with newborn kids. I was horrified to find I'd prophesied completely wrongly. These kids, and for that matter all the kids in the world, can pass the glass-table test as soon as they stand up. It had nothing to do with sight—lambs with occluded eyes showed no difference in their results from sighted animals. We had to conclude that precocial animals (i.e., animals able to run around at birth) can pass the test where others cannot.

"Then I got tired of the sheep—they're too stupid—and moved in on dogs. We all know the dog was domesticated by Homo sapiens, but most people think the reason was that man wanted a hunting companion. In fact what he wanted was a herder, and that's what he got. Dogs that haven't been taught how to herd know just what to do anyway: they'll spread out and herd animals such as sheep, and they know how—at least, they know how after they've reached the stage of sexual maturity. It's not taught behavior. I kept border collies in absolute isolation until they were twenty-one months old. You could take one of those dogs out of isolation—though you'd have to handle him with gloves—and within six hours he'd be herding. If you did the same with an Alsatian, he'd become a wolf.

"Sometimes you can see coyotes around here. You can watch them half a mile off and see that they work in pairs. The herd they go after flocks together, and then the coyotes charge. Incidentally, cosseted sheep *won't* herd: naturally they won't, because they're not afraid of the things other sheep fear. I found female border collies difficult to train because they are so aggressive. You must have a little aggression for success, but not too much: the difficulty is to get just the right amount. Well, in the end the dogs were too much of a responsibility, and I handed them over to the medical school, where they went on being troublesome because they were crazy. One of them even had hallucinations. . . .

"I got involved with primates through woolly monkeys in about 1959 when I bought a pair of them for my small daughter. They were young, and they got imprinted on us and lost interest in each other—they've never bred and never will. And they're very aggressive, though they still love my daughter and me. There's one very odd thing about woollies—they grind their teeth to keep them sharp."

We had ambled over to a field in which roamed a large number of black pigs, and he led the way to an enclosure where several litters were busily eating. He said that they were African Guinea hogs, and very valuable for research purposes. "They're line-bred and inbred, and that means they're always predictable in their reactions. But when we first got them, they didn't do at all well. They were fed the very best food and kept on nice clean concrete flooring, but they didn't do well until we let them off the concrete and slimmed down the sows. They roam around now pretty much anywhere, and they're doing fine."

Next day Dr. Lemmon and a girl assistant named Mary picked me up at the hotel, and we started off to survey some of the young chimpanzees who had, as he put it, been placed in foster homes.

"What's the object of this exercise?" I asked, "and how do you find foster parents for an odd job like this? Are they all members of the faculty like the ones you mention in your article?"

"No, not all of them. Some are people who can afford the extra time and trouble involved, and who are willing. All our foster parents are volunteers."

"You sold them a bill of goods, you mean."

"Not a bit of it," he said. "Any time one of the parents wants to give up the animal I'll take it back; they all know that. But women tend to get emotionally involved with chimpanzee babies once they start taking care of them."

I said, "Of course, if they're childless women—"

But Dr. Lemmon said it wasn't like that at all. "I look for a normal family life for my chimp infants," he said. "Preferably a place where the animal will have siblings. The first people we're going to visit, for instance, have several children—other children."

Mary leaned from the back seat and said, "Mrs. Marryat wants us to stay for lunch, she said. She was going to call you, but she tried too late."

The professor looked doubtful. "I don't know if there's time. We'll see."

The house we first entered was markedly luxurious, with archways and tiled floors and a conservatory. Our hostess, in a lacy hostess gown, greeted us with the announcement that we simply had to have a drink, and we did not protest. Several adolescent children came in to say hello.

Our hostess said, "Ephraim's being changed, but he'll be in, in a minute. How's Connie doing? We'll have to have a birthday party for the two of them."

Dr. Lemmon rose to the bait. "No species contact," he said fiercely. "You know that. If I hear of those two chimps meeting each other—"

"There, there, I didn't mean it," she said, as a black woman in an apron came in, carrying a small chimp baby, three months old, swaddled in a fluffy blue blanket.

Everyone said, "Oh, isn't he the sweetest thing?" and the children hovered over the infant, so it was a little while before I was close enough to look at his tiny black face. He was put in my lap, and there pulled himself out of the blanket and attached himself to my clothing, as young chimpanzees like to do. He was fully dressed in a warm bunny suit.

"Isn't he adorable?" asked the foster mother, and gave us all another drink. The chimp was passed from hand to hand, from chest to chest: he leaned out and felt a lily leaf with small, tentative fingers.

As we drove away after lunch, waving, Dr. Lemmon said, "They're hardly any trouble at this age. The real test comes later on, when they begin investigating the world. I never feel sure, after I've placed a chimp, how long it's going to last, but I think this foster mother has a good idea of what's in store, and I have hopes she'll last a long time.

"I'm not so sure about the setup we're going to see now. The prognosis is bad: soon after they got the chimp, the wife got pregnant, and the baby has now been born. So far it's all right—they simply hired an extra nurse to look after the chimpanzee—but as the two infants grow older, the chimp is going to be more advanced than the child; it will walk first, and all that. The question is, Can the parents stand the strain? They might grow jealous on account of the human baby."

This house stood in deep country and was even more luxurious than the first one. I left Bill Lemmon downstairs and went with Mary to see the chimpanzee of the family, who, we had been told, was in her foster mother's bedroom. There we found her sitting on the pale green carpet, playing with a Dinky car all among the crystal and satin of the furniture. She was wearing a denim-blue diaper.

"There you are, darling," said Mary, and set down the little black bag she was carrying. "Do you want to be picked up, or would you rather come by yourself? . . . I wonder if you'd mind stepping back a little—you might sit on the bed," she said in an aside to me. "She's probably wary of you because you're strange. I'm supposed to visit her every two weeks, and see how she's developing, but she's getting older now, and I usually come once a month."

The chimpanzee grew impatient of being ignored, and held up her arms to Mary who lifted her, saying to me over her head, "This little one has problems. It's not her fault, but she's had three pairs of parents, and that's bad for babies—Isn't it, darling? Now let's see what you want to play with."

Out of the black bag came a string of beads first, then a hand mirror. The chimp bounced the beads a little and looked into the mirror with a somewhat jaded gaze: evidently she had seen these things before. She ignored a small doll, but paid some attention to a bunch of jingling bells. Little by little she lost her shyness, ran behind the curtain, and played peek-a-boo, and climbed up on the bed to bounce. When the session was ended and we opened the door, she preceded us down the stairs, sometimes swinging on the banister, sometimes hopping from step to step. In the sun room, where we rejoined the party, she climbed straight into her surrogate mother's lap.

"She just loves the baby," said that lady, beaming, and I could see that the professor was relieved.

Our last call was paid on Dr. Lemmon's secretary Jane Temerlin, a psychology graduate, whose husband too is a member of the psychology department. Their house was nearly as secluded as the Lemmon residence, with a lawn and a pond in front of the veranda. As we entered the living room, Dr. Lemmon said, "Notice how chimp-proof this place is. Plastic furniture covers, washable walls, special locks on the doors—"

"You learn as you go along," Jane said. "We had to replace a lot of stuff at first, but now we know more about it. Lucy ought to be home pretty soon; she's out shopping with her father."

The door opened, and a chimpanzee about three or four years old came in, ahead of Jane's husband.

"I get more impatient with people all the time," said the foster father of Lucy, handing packages to his wife. "Whenever I park the car downtown and they see Lucy, they crowd around. That's natural enough, I suppose, but then some idiot is sure to say, 'Is that a monkey?' What the hell do they think it is?"

"Try saying no, and see what happens," Dr. Lemmon suggested.

Lucy sat on the floor next to my handbag and began with practiced fingers to open the zipper. Then she paused, and looked up. Dr. Lemmon said, "She didn't say no, Lucy," and the chimpanzee immediately turned back to the bag and opened it all the way, took everything out, one by one, and examined them. She recognized my compact at once, and opened it to powder her nose in the little mirror; then with one careful finger traced the line of her eyebrow as she watched her reflection. She opened a bottle of eyedrops, sniffed at it, and replaced the cap. When everything had been looked at, she returned the articles to the bag and closed the zipper. Seeing a smoldering cigarette-stub in an ashtray, she leaped up on the chair and tamped it down, extinguishing the spark. Jane was telling Dr. Lemmon about the additions they were having built on the house: a cage on the roof, with a caged passageway and room leading up to it from the interior. These well-locked apartments were to keep Lucy in during the day.

"We've simply got to do it," she said. "She's getting so clever with locks and doorknobs there's no holding her in the house as it is, and we both have to spend so much time at work. . . ."

Away from Oklahoma, I thought often about the colony, and Lucy, and the other adopted infants. I had been much impressed on the day of the visits by one brain-washed foster mother whom I overheard pleading with Dr. Lemmon—"You're not taking Gordon away, are you, Bill?"—but mostly I thought about Lucy, and when I returned to Norman, after a few months, I was glad to hear that all was well at Jane's house.

Before I went to see that household, however, I was to watch the Lemmon apes at work. By appointment I joined Dr. Lemmon and a graduate student named Jerry, and we drove out together to the farm. Jerry, said Dr. Lemmon, was working for his Ph.D. on a paper with the title, "Dominance in Chimpanzee Relationships."

"According to Jane Goodall's observations, dominance interactions are rare in the wild," he explained, "but at Holloman there certainly is a lot of interaction of this kind. The question we want to answer is, Is it simply a matter of less space? Jerry is trying to find out by reducing the living space of our chimpanzees still further than the space they normally have, to see what happens. Yerkes said that every animal in captivity strives for status in his group. Very likely, we thought, you'll get a linear dominance hierarchy when you reduce the space—in fact, if animals didn't have it, they would probably go under. Then if this is true, as space is again increased, the hierarchy will dissolve.

"And that is how things actually do work out in these experiments, with an added result: not only is there a constant struggle for simple dominance, but you get coalitions in the group, against the boss man. We say coalition because it's a favorite word just now among minority groups. This is what happens: Suppose you have a group of three, A, B, and C. C becomes dominant, upon which A teams up with B to form a coalition with which to overthrow C. So far we know this—really know it, that is—only of human beings and their behavior, and we also know that in this struggle human males and females are usually different. Males are aggressive, females passive. Is the same thing true of chimpanzees? Jerry is trying to work that out, combining the ideas of space effect and coalition."

"Here's how we do it," said Jerry. "First we put only two animals in a very small place to see if we got a linear hierarchy, and we found that we did: every single-pair relationship always came out like that. There was no *permanent* chief, but for the moment the relationship was fixed.

"I'd better explain our system for making sure. We have a dispensing machine on top of one of the cages in which we put grapes, pieces of banana, or anything else that chimps like. The dispenser is connected with a slot machine in the next cage. If a chimp puts a nickel (which we give him) into the slot and pulls a lever, a grape or piece of banana falls out of the dispenser into the first cage. Now, the chimp with the nickel is in a disadvantaged position—a long way from the fruit when it falls. There's no direct way from cage to cage: he's got to run out of the slot-machine cage and into the dispenser cage before he reaches the fruit, but if he's the dominant one, the other chimp will probably be afraid to take it. However, if the other chimp is dominant, he grabs the fruit right away. All we have to do is count the grapes each animal gets, and the one who gets the most is the dominant one—at that time.

"And the thing holds true even when you've got more than two chimps. Suppose you have three. If you start out with 120 grapes and one animal gets 68 of them, that leaves 52 divided between the others: say one of them gets only 18, the other has 34, which means the situation is fairly stable. But now we get the coalitions. Let's start again with three chimpanzees and 150 grapes of which the high man gets 70, and the two lower 50 and 30 respectively. Fifty and 30 add up to 80, and that's more grapes than the dominant chimp has scored. The other two make a coalition, rise against his authority, and overthrow him. To a significant level you can predict coalitions just by counting grapes: it almost always works out. Operationally we define a coalition as two against one. Well, in one period, out of thirty-five of these tests there were sixteen overt fights, or struggles, to change the coalition setup—*overt;* naturally there were also many other struggles we didn't see. But you also get another pattern. In eleven out of sixteen combinations, where we were using three-chimp groups, the top and bottom man united against the middle."

"Like the workingman uniting with the government against the businessman," said Dr. Lemmon. "We're almost ready for the second half of the experiment—to increase the chimps' space and see what happens."

Jerry added, "There are all kinds of complications. For instance, once in a while the chimp who is supposed to put the nickel in won't do it; he just refuses. That's stalemate. In the two-cage system the nickel-dropping chimp catches on very quickly that he'll never have enough time to get back before the grape drops. Withholding the nickel is a subtle manipulation. And there's another manipulation they've figured out: two or three of the ones who aren't putting in the nickel start romping around and playing, hoping to take Number One's mind off that grape just before it drops. In the two-cage system the hierarchy tends to fall apart in any case because the situation's so ambiguous: when we were running tests in just one cage, we had only seven coalitions as compared with some fourteen so far in the two cages. There's less overt fighting in two cages—aggression tends to drop out. They prefer winning in more subtle ways."

Arrived at the Lemmons' house, we went straight to the cages. Two of the assistants came in with us, and to the accompaniment of tremendous noise they lifted some doors and lowered others, shooing indignant apes this way and that, until they had one male, Pan, and two females named Mona and Wendy isolated in the two cells that had been described to me, each with a door to the side, so that to move from one cell to the other a chimpanzee had to detour. A flat black machine hung broodingly over the nearer cage, and on the floor of the other stood the slot machine.

The tests began when all three apes were in the front cage, with the fruit machine above their heads. One of the assistants handed a nickel to Wendy, and Dr. Lemmon said, "Okay, Wendy, put it in the slot."

Shamblingly, reluctant to leave the two others, Wendy obeyed—out through the front cage door and in through the rear cage door, where she put the nickel in the proper slot. For a second or two nothing happened, except that Wendy started running back as fast as she could.

Then, with a wheeze and a clank, the fruit machine dropped a grape into the cage, right into Pan's hand. He swallowed it.

"That's the way it usually turns out with these three," Jerry said.

"But sometimes the girls go on strike," said Lemmon. "Once and only once Wendy was dominant for a little while. It didn't last long, though."

The third chimpanzee, Mona, had made no attempt to get anywhere near the source of the fruit, but remained hanging on the bars, which she clutched with hands and feet, paying no attention to the game. Again Wendy put in a nickel and raced back, and again Pan calmly grabbed the prize.

The professor said, "Mona completely ignores the power struggle, you see, and doesn't even try for a grape. She's detached except in special circumstances. We'll give her a chance later on."

Two or three times more the grape was picked up by Pan, until the men decided that we ought to have a change. To the accompaniment of more screaming and shooing, Wendy was removed, and another chimpanzee named Mimi was put in. Mona became the one who put the nickel in, but there was a new development when Pan, perhaps lulled into carelessness, didn't grab quite as quickly as he had been doing and Mona bustled back and nearly got it. Pan saw her in the nick of time. On the next occasion she had actually picked up the grape before he started for her in a threatening manner, when she dropped it and ran away.

"Both those girls are scared to death of Pan," said Dr. Lemmon, "that is, nearly always they are. Coalitions seldom work against him, and they don't even try to overthrow him: they're licked before they start. But sometimes among the lower echelons you see coalitions, even when Pan's running the show."

Now Pan and Mona were removed, and two new females appeared to work for a time with Mimi—Carrie and Peggy. The men told me in advance that Carrie was stupid, but that Peggy, only half her age, was smart. It was Peggy who had first figured out that if she got up close to the hole through which the fruit came she would get it. "She's the only one who even tries to manipulate Pan," said Dr. Lemmon. "But don't sell Mimi short, either— she's smart, socially. This is a difficult combination: each tries to outwit the others, and the pattern's never distinct."

A nickel was handed to Peggy and she went slowly at the task of dropping it in. Suddenly the two others, Mimi and Carrie, began wrestling with each other on the floor just under the dispenser. Such struggles, such ticklings and gasping laughter! They never, however, rolled away from that spot on the ground that was under the machine, and Peggy stopped short to watch before dropping the nickel.

"Peggy's not suckered, not for a minute," said Bill. But she dropped the nickel at last; the grape fell, and Mimi grabbed it.

"Poor dumb Carrie," Jerry said. "She never wins." As for Peggy, she sulked.

I was told that Mimi once kept a nickel hidden in her mouth all weekend, from Friday to Monday, when the game wasn't played, until she saw her chance to put it into the machine and get the reward. "She *planned* it," Dr. Lemmon said proudly. Suddenly he roared, "Peggy, leave that apparatus alone!" Turning to me again, he explained: "She's learned how to break it. If she puts the nickel in a little crooked, the whole thing stops, and when she's had a losing streak, she's apt to do it." He grew pensive. "It's funny about money and all that. Once Mimi actually took the nickel away from the three other girls, Carrie, Mona, and Peggy. She didn't take food, mind you, but money. Pan will sometimes take food out of the others' mouths, but even he has never committed robbery like that."

One evening after work Jane drove me out to her house for dinner, confessing on the way that she was a little apprehensive about what we would find when we got there. Lucy's cage on the roof was finished at last. "It's the first full day I've left her alone the whole day in her new apartment, and I hated to do it," she said. "But you can see how it is: I didn't have any other course open. Our son's away at school practically as long as we're at work. He's sixteen, and I can't expect him to play nursemaid."

As we got near the house she said, "I don't see her up there, do you?" We looked hard, but saw no sign of Lucy. "I expect she heard the car, and she's waiting in the downstairs part of the cage where we keep her toys," Jane said, but we hurried to go into the house. There was Lucy behind the bars, safe behind the padlocked gate, sitting in a basket and rocking busily back and forth. She gave us only a fleeting glance.

"Hello, darling," said Jane. "You're mad at us, aren't you? Rocking like that, just to make me feel worse." She unlocked the gate. Lucy got out of her basket and ran past Jane, arms outstretched to embrace me. "She's punishing me," said Jane. "I wonder how long she's going to sulk."

Lucy sulked nearly all evening. She ignored Jane, who was cooking in the kitchen, though Jane's husband and son got big welcomes when they arrived, and so did Dr. Lemmon. While we had cocktails, the chimpanzee played with me, teased the men, and had a wrestling match with her surrogate brother. She sat at table with us and ate her dinner, but continued to ignore Jane until she snatched at a wine glass and Jane scolded her. Then, almost quicker than the eye could follow, the chimpanzee grabbed up a piece of pineapple and shoved it into my mouth. The gesture seemed to mark the end of the feud; a few minutes later she fell asleep on Jane's lap.

"It's a strange state of mind one gets into," Jane said, "as if one had a retarded child or something like that—I think it must be exactly like that. I get so defensive: I know it's silly, but there it is. Yesterday I bought food for Lucy at the store down the road, two dozen cans of Gerber's Baby Food, and the man wrapping them up said, 'This is for your monkey, eh?' And I said very coldly, 'No, it's for my daughter.' "

Recently I heard from Jane. She wrote that the Gardners, who have a chimpanzee, Washoe, whom they are training to use a sign language, had just been to Norman on a visit, and had been delighted with "the slender, flexible fingers" of most of the young Norman chimpanzees. Washoe's fingers are short and stubby, they said: it might be easier to teach gestures to the Norman apes. Later they sent Lucy several of the kind of dolls that Washoe likes best.

"On [Lucy's] introduction to the largest of these babies [about eight inches high, very round and chubby]," wrote Jane, "she exhibited all of the appropriate adult chimpanzee maternal behavior segments. She carried it properly, gently, . . . when it 'cried,' she reacted just as our good mothers do—stopped, looked at it, readjusted it, groomed it, put her finger on her nose, then the baby's nose, ear, etc. All of these things and more were chimpanzee appropriate, not human, even though her maternal care has been exclusively human. It seemed clear that if this had been a live, responsive, clasping chimpanzee baby, Lucy's now available repertoire would have meshed with that of the infant to establish 'good' maternal behavior."

11

Speak No Evil

Ever since men began talking, they have told each other stories of other talking animals, like those in *Aesop's Fables* or Balaam's ass. A Portuguese explorer named Damiao de Gois, who lived in the sixteenth century, wrote of an Indian elephant he had seen that could inscribe legible letters on the ground with the tip of his trunk and, when asked what he wanted to eat, replied loud and clear, "Rice and betel nut." De Gois' contemporaries believed his story. We might almost believe it ourselves, since we have such a high opinion of elephant intelligence, but our notions of intelligence tend to be subjective and we are not clear-headed about it. For instance, are we sure that the power of speech is a sign of intelligence? Or, does it go without saying that everyone intelligent can speak? Is speech, in fact, intelligence itself? Some people have gone so far as to say so. St. Augustine didn't, quite, though he said it is a mystery that we men should be able to articulate meaningful words; but Descartes said outright that the difference between man and the other animals is that man alone is capable of abstract thought and has language. He concluded that a man's "body machine" is somehow attached to his "thinking machine," that language and abstract thought are interlocked, and that the infrahuman animal, *because* he is unable to speak, is incapable of thought. The idea of interlocked thought and language is still with us: the noted Swiss psychologist Jean Piaget says that logic and language are obviously interdependent.

Language and speech: their definitions have plagued wise men ever since Babel was abandoned. What is speech? How do we learn it? How do we produce, how receive it? If, as some aver, infrahuman animals are not talking when they make noises, what is it they are doing? Many animals utter sounds, or—to use the scientific terms—vocalize or phonate, and the sound of each species is characteristic of it. Dogs bark, cats mew, babies cry. The dog is broadcasting a message, perhaps a threat or a demand for attention. The cat reminds its master that it is hungry or needs a pat of reassurance. The crying baby signals that it is wet, or frightened, or hungry, or bored; but none of these creatures is talking as we talk. The baby is simply too young to know how; the dog and cat, no matter how old they grow, will never talk in our speech. Here and there a dog can be taught to make sounds approximating human speech, but his utterances are mechanical tricks, fit for the circus rather than the *conversazione*.

Looking at the animals most closely related to us physiologically—the anthropoid apes—observers often wonder why they shouldn't be able to use their vocal cords, lips, and tongues, which look very much like ours, as we do. Three of the four great apes (as usual gibbons are special, and can't be included for our purposes), the gorilla, the chimpanzee, and the orang-utan, really are so similar to us that we fall naturally into thinking of them as children or slightly subnormal or retarded human beings, and hopeful people have tried to bridge the language gap by bringing the gift of speech to them as her teachers handed it to Helen Keller.

Robert Mearns Yerkes gave considerable thought to the subject. After all, he had studied Wolfgang Köhler's work and knew how chimpanzees can solve puzzles: he himself worked with at least one orang-utan and one gorilla, as well as chimpanzees. The impression he got was that the three species are all quite surprisingly intelligent, each in its own way, and it was hard not to assume that this intelligence might imaginably lead to the faculty of speech if proper lessons were given.

It was a natural assumption—today it is still natural, unless one accepts that puzzle-solving intelligence is not quite what is needed for speech. As for intelligence, there can be no doubt of it. There is a training-school for chimpanzees at the London Zoological Society's headquarters in Regent's Park, where at a certain time of day the public may go into the Children's Zoo and watch classes. There young chimps learn to fit odd-shaped pieces of wood into matching holes, as human children do in nursery school. They match colors, figure out how to extract bananas from tight-fitting pipes, and capture fruit, hung high out of reach, by piling up boxes to stand on. Under the alert gaze of a young ape, the trainer opens the door of a safe furnished with bolts, hooks, and various locks and keys, and inside deposits a grape or some other tidbit that young chimps like. He closes and locks the safe with all its devices, then steps back for the chimpanzee, who quickly undoes all the locks and bolts in reverse order, without mistake, until the door swings open and he grabs the reward. Even if one dismisses this display as a mere feat of imitation, the fact remains that it is imitation in reverse, which is remarkable.

Often it is difficult to draw the line between imitation and creative thought. My friend Chimpo was usually kept confined by a chain around her neck that was fastened with a small padlock. Most of the time I kept the other end of the chain hooked around some strong, heavy object, so that she was tethered—it was a situation she disliked intensely. At other times, when I could keep an eye on her, she was less confined, the chain unconnected except at her neck, but acting as a sort of brake. If necessary I could grab it as she went past. That situation, too, she did not like very much, though she probably preferred it to outright captivity. On rare occasions, when nobody else was around to be bothered by unruly apes, I unlocked the padlock and let her go altogether free. This she did like. One day I found her sitting on the floor near the tied-up end of her chain, deeply engrossed in a project so astonishing that it was years before I spoke of it to anybody for fear of being considered a nut. She had got hold of the door key—a large, heavy piece of metal—and was trying to insert it into the keyhole of the padlock at her neck. She was clumsy at it, and the door key was ten times the right size anyway, but there could be no doubt that she had grasped the *concept* of keys. If you had asked me then what her chances were of learning to talk, I would have replied emphatically that it was only a matter of time.

The far more knowledgeable Yerkes had this idea, and wondered why apes don't talk. To be sure, chimpanzees are far from silent as it is. But there are differences, he pointed out, between vocalization, speech, and language. He said they are three separate things, though they combine. "Vocal sounds are not necessarily speech," he wrote in *Almost Human.* Speech is many things, including vocalization, but vocal sounds may be merely simple, monotonous cries that signify emotional crises. Of language, which includes speech, there are many sorts: some, such as sign language and writing, are not vocal. Though man evidently has a "special gift" for inventing and developing spoken and written language, it would be rash to assert that he alone has developed language. Consider the ants, Yerkes told his readers: they communicate with each other though they don't talk. Ants have a "language."

He made himself familiar with the writings of Richard L. Garner on speech among primates—three books, published in 1892, 1896, and 1900 respectively. Garner had no scientific training, but he had a hobby. He spent much time close to monkey cages in various American zoos, listening to and recording the noises made by the inhabitants. At first much of this activity depended on the goodwill of zoo authorities and keepers, but after he published his first work, interested readers urged him to go to the wild and make more recordings, and he did go, to West Africa, where he built a cage, lived in it, and observed, or listened to, primate life. His methods and conclusions alike were ridiculed by most of the contemporary scientists, but Yerkes paid attention, made allowances, and believed part of the writing.

Garner singled out special sounds that seemed to him to be used repetitively by his simian subjects, in certain situations to which they seemed consistently connected—when they were being fed, for example, or warning one another. He claimed that these sounds were monkey "words," and he believed it quite logical—as, in a way, it is—that a Capuchin monkey should speak one language and use a certain set of words, while a spider monkey has a different language and different words. Even so, certain words seemed to him common to both species. Garner tended to generalize from the particular: when he saw a monkey that shook its head with what seemed to be rejection, he deduced that such a sign for no is universal among primates, including man. These ideas, however, were not what interested Yerkes most. In Africa Garner paid much attention to chimpanzees, recording a chimp vocabulary of what he claimed were twenty-five to thirty words. Then he set about teaching an infant chimpanzee, Moses, to talk our language. He claimed that after some time the animal was able to say four human words: "mama," *"feu"* (French for fire), the German *"wie,"* and the Nkami word for "mother," which Garner spelled *nkgwe.* (I know nothing about Nkami, which is evidently an African language, but if Moses managed to utter *nkgwe* he was an outstanding primate.)

Garner described his teaching method, and the description is worth repeating because it has served as a model for at least two experimenters since then. Every day he took the little ape on his lap and coaxed him to imitate noises he himself made, repeating the four words again and again. For a long time during this exercise the chimp watched the man's face but did not react. At last, after many lessons and much reinforcement, Moses began "dimly" to see what was required of him. Trying to say "mama," he worked his lips soundlessly. He managed *"feu"* better, though it came out sounding more like "veu," and he had trouble with the initial *w* in *"wie."*

So far, the amateur. In 1909, in a scientific journal published from a psychology clinic in Philadelphia, Lightner Witmer reported that a trained chimpanzee named Peter could, though with difficulty, pronounce the word "mama," and that he evidently understood its meaning. There was little noise in his pronunciation, which was breathy. That was all there was about Peter, but on April 13, 1916, Garner's feat with Moses was brought back to mind in a paper read by Dr. William H. Furness 3d —who was just as respectable as he sounds—to a meeting, in Philadelphia, of the American Philosophical Society. Dr. Furness explained that he had spent some time in Borneo, there came into contact with orang-utans, and became "possessed with the idea that with constant human companionship and surroundings at an early age, these anthropoid apes . . . were capable of being developed to a grade of human understanding perhaps only a step below the level of the most primitive type of human being inhabiting the island. . . . If deaf, dumb and blind children have been taught by beings they could not see to use language they could not hear would one not be justified in an earnest endeavor to teach the higher apes with faculties and senses alert and with traditional powers of imitation, to do the same to a limited degree?" It seemed well nigh incredible, he continued, that in animals that are otherwise so close to us physically there shouldn't be a rudimentary speech center in the brain that only needed development. But though he had made that earnest endeavor, and was still trying, he couldn't say that he was encouraged.

In the course of the experiment he had acquired two chimpanzees and two orang-utans, and one of each species took part in his experiments. He had spent up to six hours a day for weeks at a time working on the problem, but that, he said, was not a hundredth part enough to teach them "articulate speech." Neither orang nor chimp used his lips or tongue in making natural emotional cries. Dr. Furness startled me by saying (or writing) quite calmly, with only a touch of exasperation, "In the case of the orang-utan it took at least six months of daily training to teach her to say 'Papa.'" Think of it: at least *six months.*

He said he had selected this word "not only because it has a very primitive sound, but also because it combined two elements of vocalization to which orang-utans and chimpanzees are, as I have said, unaccustomed, namely: the use of lips and an expired vowel sound." He had repeated the sounds for minutes at a time while manually closing and opening the ape's lips in imitation of his own. Sometimes he arranged a mirror to reflect teacher and pupil, so the pupil could see what her lips were supposed to be doing. Suddenly—even if not suddenly enough to suit Dr. Furness—it happened: ". . . one day of her own accord, out of lesson time, she said 'Papa' quite distinctly and repeated it on command." Afterward she never forgot it, and there came a time when she recognized it as Dr. Furness' name. When asked "Where is Papa?" she would point to him or pat him on the shoulder. This part, at least, sounds convincing: orangs do pat their friends on the shoulder, as if in affection or approval. One summer day Dr. Furness carried his prize pupil into a swimming pool. "She was alarmed at first," he said, "but when the water came up to her legs she was panic stricken; she clung with her arms about my neck, kissed me again and again and kept saying 'Papa! Papa! Papa!' Of course I went no further after that pathetic appeal."

The next word he taught her—possibly because he understood the mechanism of the human mouth in producing it—was "cup." He pushed her tongue back in her throat so that she could make the sound "ka," with a spatula pressed lightly on the center of the tongue. When she had taken a full breath, he put his finger over her nostrils to make her breathe out through the mouth. "The spatula was quickly withdrawn and inevitably she made the sound 'ka,' " said the doctor. After several days' practice the orang would draw her tongue back without his having to put the spatula into her mouth, but she wouldn't say "ka" unless the doctor put his finger over her nose as he had first done—until she found that her own finger would do as well. After that, her finger on her nose, she said "ka" whenever her papa asked her to, and from that point on, he says, it was easy to teach her to say "cup." He simply closed her lips as she came to the end of "ka." She knew what this word meant from the start. Once when she was ill, she leaned out of her hammock and said, "Cup, cup, cup." Sure enough, when given a drink, she proved very thirsty. Dr. Furness was teaching her the diphthong "th" when suddenly she died.

Summing up, Dr. Furness was gloomy as to chances of educating chimpanzees and orang-utans any farther. He said he had perceived only the faintest rays of evidence that they were capable of reasoning, i.e., deducing an inference from certain premises, "unless association of ideas, which in point of fact is merely learning by experience, is reasoning." For instance, "the chimpanzee if given the key to the closet in her room will fit it in the lock, turn it in the right direction, slip back the little spring catch, open the door, get to the top of the spigot which is kept there to avoid a waste of water, fit the top of the spigot, get a drink of water, and finally turn the water off." Dr. Furness was dissatisfied with this; he considered the act governed by what he called a simple succession of ideas rather than a prearranged sequence of actions with a definite object in view. No doubt he knew what he was talking about, but *I've* known chimps who wouldn't have turned the water off. He concluded that apes' brains are as incapable of reasoning in human fashion as a dog's paw is incapable of holding a pen, man fashion. ". . . their highest notch of mentality after four or five years of training is hardly comparable to that of a human child of a year and a half," he said scornfully.

Yerkes read the report carefully, as he had read Garner and similar literature. At last in 1923, spurred on by such examples, he decided to try a few experiments of his own along the same lines. He had recently acquired the two young chimpanzees, Chim and Panzee, of whom he was to make a comparative study, and he resolved to include in the study attempts to teach Chim—not Panzee, who was sickly and uncooperative—to talk. Out of his experiences came the book *Chimpanzee Intelligence and Its Vocal Expressions,* written in collaboration with Mrs. Blanche W. Learned, a friend who interested herself in music. Among the conclusions expressed in the book is the statement that there is slight if any tendency on the part of chimpanzees to imitate sounds, though his animals imitated many of his acts. Not that they were dumb. Both Chim and Panzee had "excellent" voices and the ability to produce a wide range and great variety of sounds, though they exhibited only a few types of vocal reaction. And certain of the sounds they made were characteristic of certain situations. In other words, though Yerkes did not say so, Garner may have had the right idea, for the chimps would make one particular noise consistently for situations or objects desired or liked, and quite another for objects or situations disliked, resented, avoided, or feared. To anyone familiar with these noises, the chimpanzee's state of mind or desire might easily be interpreted. But Yerkes found it impossible to take Chim beyond this mode of self-expression.

"Four methods of speech instruction have been tried," he wrote after a period of eight months, "and each in turn abandoned because of lack of positive results." First his attendants helped him rig up a chute leading from a hole in the observation-room wall down to a table in the room. From time to time the experimenter outside went to the hole and said "bă, bă" very clearly and impressively, at the same time shoving pieces of banana down the chute to the eager little Chim. This was done once or twice a day for two weeks, and during that time Chim never once tried to say "bă, bă" in reply. He didn't try to say anything. Brightly interested at first, he soon lost his verve and cared only for the banana bits.

Yerkes abandoned the chute. He next tried a box that could be loaded with pieces of banana that were released by a spring, and hung the contraption on the wall where Chim could see it. Each time he came up to examine it closer, the experimenter said, again distinctly and emphatically, "cō, cō" and released a piece of banana. Once in a while the experimenter grabbed the banana ahead of Chim and ate it, to keep the chimp stirred up and keen. "Occasionally this procedure induced certain lip movements seemingly in imitation of those of the experimenter," wrote Yerkes, and once in a great while, possibly by accident, Chim made a sound. It seemed to the observer that he was really trying to utter, though his attempts came to nothing. But that was all, and after a few weeks the box test, like the chute, was laid aside. The third trial involved another box, a small one containing—of course—bananas, with a wire-mesh cover: as the experimenter said "nă, nă," a banana was uncovered. Yerkes got absolutely nothing out of this, and gave up.

Mrs. Learned's contribution was a record of the chimpanzees' sounds in musical notation, and a chimpanzee's diary to show when the different noises were uttered. Both animals made what she called "the food sound" when they saw their meals approaching and sometimes while eating, as if they were commenting approvingly on the food. *Gahk* was the root word for food, she thought, but she admitted that it had several variants—*ngahk, ghak, khak, gak,* and many more; even *gah* as in *gah ah ah gah gah ack ack ack ack*. She also noted that chimpanzees are capable of vocalizing simultaneously on two pitches, usually at moments of excitement or fear.

At Yale there is one particular anecdote about Yerkes. A visitor once commented enthusiastically on the black woman who helped take care of the chimpanzees. "Charming!" said the visitor. "How nice for you to have a real old-fashioned Southern mammy for this work. I'm sure she's a great comfort to you."

"Mm-mm, yes," said Yerkes, not quite wholeheartedly. He added with asperity, *"But she has her favorites."*

In 1927, Winthrop N. Kellogg, associate professor in psychology at Indiana University, decided to try to humanize a chimpanzee for scientific purposes. He thought the experiment might shed new light on the classic environment-heredity controversy. Recalling stories of "the wild boy of Aveyron," who spent his early years as an animal among animals, and the similarly circumstanced wolf children of India, all three of whom never succeeded in adapting themselves to civilization after they were rescued, he wondered if environment might not prove to be more powerful than heredity in a similar situation, carefully observed. Since Kellogg could not hope to duplicate the wolf children's experience by subjecting a human child to a similar experiment, he felt that the next best thing would be the creation of the opposite situation—a wild infant brought up in civilization. Wild human children were remarkably hard to come by, so he proposed using an infant chimpanzee instead. Today his overall plan seems full of holes, but the idea was interesting.

It soon became clear that even infant chimpanzees were not found under every gooseberry bush. The professor was still looking for one when Mrs. Kellogg—who had never been quite so keen on the chimpanzee idea as her husband was—gave birth to their only child Donald on August 31, 1930. What followed is recounted in a book written by the Kelloggs, *The Ape and the Child: A Study of Environmental Influence upon Early Behavior*. Donald's birth did not appear to his father to be an obstacle. The chimpanzee was to be brought up absolutely like a human child in every respect: What could be more natural than an elder brother in the family? At last, on June 26, 1931, thanks to the help of Dr. Yerkes, the Kelloggs collected and took home a female chimpanzee who was released on special loan from the Anthropoid Experiment Station at Orange Park, Florida. Her name was Gua. She was born in the Abreu colony in Cuba on November 15, 1930, which made her two and a half months younger than her foster brother: she was taken from her mother at seven and a half months. Kellogg reflected that it was a pity she was not younger, since early impressions are supposed to be most important, but it couldn't be helped.

Gua became in every way possible a typical middle-class American baby. She had her own cot and high chair; she wore clothing; and she followed a strict, full routine. So did Donald. She and the boy were together almost every minute of the waking day. The Kelloggs kept parallel records of their development: they were careful to maintain detachment as far as they could, and the tone of the records, like that of the book, sometimes becomes almost ludicrously impersonal, as in the chapter "Some Basic Similarities":

> The difference between the skulls can be audibly detected by tapping them with a bowl of a spoon or with some similar object. The sound made by Donald's head during the early months is somewhat in the nature of a dull thud, while that obtained from Gua's is harsher, like the crack of a mallet upon a wooden croquet or bowling ball.

Other differences between the infants soon became apparent. Gua climbed better and earlier than Donald. Trying to compete with her, Donald became a better climber than most babies his age, but he could never match the agility with which Gua scrambled up into her high chair. He was better than she when it came to building with blocks. It wasn't that she failed to put one block on top of another as well as he did, but chimp hands and wrists do not bend backward, and when it came to releasing her block, Gua's long curved fingers often scooped it off again, though this wasn't her intention. In walking, too, their records diverged. Gua was already walking on all fours when she arrived. Later she adopted an upright position most of the time, and walked pretty well like that, but Donald was still shaky on his feet long after she became sure of hers. She was very good at the broad jump and often practiced jumping from different angles, whereas Donald didn't—indeed, couldn't—jump anything like as well. Both babies loved being pulled in a wagon or taken for rides in the family car. Gua was more ticklish than Donald: the Kelloggs could make her chuckle with silent, gasping laughter by tickling her or, later, merely threatening to do so. Sometimes she actually tickled herself, laughing hard as she did it. Donald was better than Gua at concentrating. If a parcel was unwrapped before the ape and the child, the ape seldom waited to see what was in it, but the child did.

They played together with such fascinated interest that often one of them couldn't be induced to eat because the other was on the floor playing, and he wanted to join in. Some games they seemed to know without being taught, like follow-my-leader or tag. But Gua enjoyed some pastimes that didn't interest Donald, when she would adorn herself with a blanket or an article of clothing and drag it around behind her like a train, or make extempore stoles or shawls with trailing moss or plants. She also loved to be swung by the hands, and Donald didn't. He imitated her better than she imitated him.

The Kelloggs described the situation as being like that of a two-child family, with Gua—though in fact she was younger—being more mature and agile, and in the role of the elder child. It was Gua who found out how to remove cushions from the davenport, put them on the floor, and jump on them. Donald merely followed suit. He had been walking for some time when he saw Gua drop to the all-fours position as she occasionally did: immediately he did it, too. In fact, his talent for imitation finally went too far for his parents' peace of mind. Let them tell about it: "His capacities astonished even those closest to him when it became apparent that he was also vocally imitating his playmate. Such behavior was first observed during his fourteenth month in the reproduction of her 'food bark' or grunt."

As Yerkes and Mrs. Learned noted, a chimpanzee makes a particular noise at the sight of food. Dr. Furness commented on the same thing: "Contentment over food seems to be expressed by grunts very much like a young pig." At such moments Gua said something like "uhu uhu, uhu uhu." My friend Chimpo said "oogh, oogh, oogh," or "oogh oogh, oogh oogh." It must be remembered, however, that we are quoting different chimpanzees. What Donald said, according to his parents, was "uha, uha, uha," or sometimes "uhuh, uhuh, uhuh."

All this time Gua made no more effort than Chim had done to enlarge her vocabulary; she seemed to rely on what the Kelloggs called a rudimentary, nonvocal form of communication—a language of action. She often used it. For instance, at mealtimes when she was hungry, she would bend her head forward so that the attendant could the more easily put on her bib, and when she had eaten enough, she herself pulled off the bib before climbing down. If she was hungry and saw a cup of milk being carried toward her, she would protrude her lips long before it arrived, or reach out and pull it closer. If she didn't want any more milk, she turned her back or pushed the cup away. When she felt like a romp, i.e., a swinging whirl, she would grab the hands of the nearest friend, walk a few steps, and then hang or swing.

She was thirteen months old when she startled the Kelloggs with an action inspired by a bottle of orange juice that stood on the floor. Gua loved orange juice, and if a bottle of it was placed in her hands in the right position, tipping toward her mouth, she could drink it very adeptly. For some reason, however, she found it impossible to lift the bottle by herself. On this occasion she tried several ways to get at the juice without lifting the bottle, and at last "she reached with her right hand toward the attendant observer . . . took the fingers of the observer's hand in her own and *pulled his hand to the base of the bottle.*"

Gua may have felt that with such powers of self-expression she didn't need speech, but the Kelloggs found it hard to be light-hearted about the situation. The embarrassing fact was that Donald wasn't learning to talk either. At the age of eleven and a half months he had given no cause for alarm; he had three words—"Gya" for Gua, "din-din" for food or dinner, and "daddy"—and according to the books a total of three words was just about right for his age. But after that he made no progress—or almost none. Things happened, all right; he added three more words to his repertoire, but with each addition he seemed to forget an old one, so that three remained the upper limit. At seventeen months he could make two sounds that apparently had meaning—an affirmative hum with an upward lilt, and a negative hum that turned down. He could also point when he wanted something, but in spite of these signs, there came a time when the Kelloggs admitted to themselves that in language their child was considerably retarded. He *could* outdo Gua in sheer vocalization, if that was any comfort: he babbled as if he were practicing, cooing or singing meaningless syllables, whereas Gua never tried to use her lips, tongue, teeth, or mouth cavity to produce similar sounds. "There were no 'random' noises to compare with the baby's prattle. . . . On the whole, it may be said that she never vocalized without definite provocation."

For several months Professor Kellogg tried to teach Gua to say "papa." He would put her across his knees in what was becoming the classic position, face upward, and say very slowly and clearly, "Pa-pa." Gua, clearly interested in the funny faces he was making, watched raptly but made no attempt to imitate his grimaces or copy the noises. Kellogg then tried manipulating her lips as he spoke, and within a month or two of such efforts, the observers were "greatly encouraged" to notice occasional voluntary lip reactions on Gua's part, as if in response to his, though it was impossible to be sure. She didn't go so far as to open and close her mouth: she merely made incidental movements, little twitches of the lips. Occasionally when Kellogg patiently repeated "papa" over and over, Gua slowly stretched out her index finger to push it into his mouth, and once or twice she followed up this gesture by touching her own lips or putting her finger between her teeth. Beyond such tiny signs there was nothing significant.

The Kelloggs decided to give up. They kept open minds, saying that their failure with Gua didn't go to show that nobody would ever persuade chimpanzees to talk: nevertheless, they ventured to predict that no anthropoid ape would ever be able to say more than half a dozen words, if that. Gua, silent to the last, left the Kellogg household March 28, 1932, at the age of sixteen and a half months, to be returned "by a gradual rehabilitating process" to the Anthropoid Station at Orange Park. The Kelloggs do not tell us what happened to Donald: presumably he made new friends and soon caught up on his vocabulary. He travels the fastest who travels alone.

Nearly twenty years went by before the world heard of another experiment in humanizing a chimpanzee. This time the book was *The Ape in Our House* by Cathy Hayes. Where the Kellogg report was carefully detached, Mrs. Hayes made no attempt to squelch her emotions. She left pure science to her husband, a research psychologist at the Yerkes Laboratories of Primate Biology—same place, new name—at Orange Park, and for herself was perfectly willing to take the personal viewpoint. The Hayeses started their experiment with an idea somewhat similar to that of the Kelloggs: they wanted to watch a chimpanzee growing up in a human world to see how it might adapt, but they placed more emphasis on the possibility of teaching such an animal to talk. The intervening years had made a difference in the supply of available chimpanzees, and Dr. Yerkes, obliging as ever, helped them to select a three-day-old animal, a female named Viki. The Hayeses moved into a house near the Laboratories and got ready. In another way their experiment was to differ from that of the Kelloggs: they had no (other) child.

Viki looked during her first days, said Mrs. Hayes, like a monstrous spider. She looked terrified and resentful and starved, her eyes staring out of her wrinkled little face and her right arm sweeping up in an arc, over and over, "searching for a mamma to cling to." She never cried, as a human baby would have done. Now and then she gave a little "oo oo, oo oo," that sometimes burst into a scream, but she didn't cry or whimper. To accustom her ear to the sound of man's world they played—constantly—a little radio near her bed. She grew older and slept less, but for a long time she clung like a burr to her surrogate mother, screaming whenever Mrs. Hayes detached her. Then, little by little, she released Mrs. Hayes now and then to explore the world.

Until she was fourteen months old, Viki made only chimpanzee noises and not many of those. Unaccustomed to human children, the Hayeses might not have noted this fact, but one day when the ape was nine months old she played with an eighteen-month-old child, Tommy, who had been raised much as she had, with the same toys, teaching methods, and general training. That afternoon, Mrs. Hayes noticed, the boy Tommy sat motionless, chattering incessantly, while Viki ran and jumped and climbed just as incessantly, but without making the slightest vocalization.

At the age of one, said her fond foster mother, she was exceptionally beautiful in a chimpanzee sort of way, and buoyantly healthy. "In physical capabilities and neuromuscular control, she was definitely ahead of the child her age." This was quite natural, for chimps mature at a faster rate than humans. Besides, "In the use and understanding of language she showed no serious deficit as yet. We did not learn until later that even at one year the human is receiving a great deal of language education, which will enable him by the age of three to rise far above the most intelligent of chimpanzees."

Otherwise Viki's intelligence was obvious. She knew the connection between flipping a switch and the lighting of a connected lamp. When the switch was flipped, she looked expectantly at the lamp. (Often it was she herself who switched it on.) At this stage she became very interested in writing, scribbling, drawing—whatever it can be called—with anything handy, pencil or screwdriver.

She got naughty and bit Mrs. Hayes. She was hard to handle. After an especially turbulent day with her, the discouraged Mrs. Hayes said to her husband Keith, "Maybe she's not as intelligent as we think she is."

She describes how they sat there sadly, silent in the gathering gloom, until Viki came in. She walked over and looked up with "soft, questioning eyes," then climbed into her mother's lap and snuggled down, pulling Mrs. Hayes's arms around her. She got up and switched on the lamp next to the chair, carried her hurdy-gurdy to the top of the bookcase, and cranked out a tune. In the distance a freight train whistled: Viki hooted a soft reply. She lowered herself to the floor and rummaged through her small suitcase, coming up with a pencil and notebook. "With these she climbed back into my lap, where she sat scribbling contentedly. Keith and I avoided each other's eyes and we never again discussed the frightful thing I had said."

The main object of the exercise—to teach Viki to talk—was not forgotten. The Hayeses faced the important fact that apes don't babble much. A human baby begins to make vocal noises at three months, and from then on chatters almost constantly. At five months a normal human infant often repeats syllables, though as far as one can ascertain they have no meaning; but apes are different. As a very young infant Viki made occasional sounds when the Hayeses spoke to her, and once in a while she seemed to enjoy playing with these noises by herself, but suddenly, at the age of four months, these bits of chatter dried up almost completely, declining in number and finally disappearing, just at the approximate age at which human children babble more than ever.

At this point the foster parents decided that they would have to teach Viki to speak by the same methods that are used for speech-handicapped children. At schools for such children the teacher manipulates the child's mouth and demonstrates the exhalation of breath, much as Dr. Furness did with his orang-utan, or in another method, likewise reminiscent of the Furness methods, the child watches himself with his teacher in a mirror and copies what he sees. But Viki had become so silent that even her old vocalizations were rare, and the Hayeses reasoned that they must start her up again, so that there would be some voice to use when it came to shaping words. Therefore, when she was five months old, Mrs. Hayes tried to teach her to "speak" for her food. For five weeks both parents worked at this task, until at last Viki did say something—"a rasping, tortured 'ahhhhh.' " From then on she would make this sound whenever they told her to speak. "It was like someone whispering 'ah' as loudly as possible and with great effort."

This was a considerable triumph, but it left questions unanswered. Why was it so hard for Viki to make that small, tortured sound when she could scream or bark loudly enough on other occasions? The Hayeses concluded that she must have been unable to make any sound whatever voluntarily. Her chimpanzee noises were merely reflex expressions of her feelings, and as such were beyond her control. Except for such reflexes, Viki had been voiceless. Mrs. Hayes concluded tentatively that the ape, for all its humanlike vocal apparatus, lacks the neural control necessary for voluntary speech. Her lack of babbling might be significant. In any case, it was plain that she had lacked the motor skill of vocalization, and now that she had mastered the asking sound of "ahhhhh," she used it in many other asking situations besides those connected with food: with it she asked to be taken up in the morning, to be taken off the potty, to be given something out of reach, or for some particular toy.

Having got this far, the Hayeses thought it time to give Viki some words, and they started with "mama"—because, says Mrs. Hayes, it is a very primitive sound, frequently a child's first word, and an almost universal designation for the female parent. Also because it would be easy to press the ape's lips together for the *m*. Dr. Hayes began the training when Viki was at breakfast. Holding her on his lap, he circled her head with one hand, thumb on her upper lip and the other fingers cradling her chin, so that he could open and close her lips. Holding a piece of food or a cup of milk in the other hand, he told her to speak, and as she made the 'ahhh' sound he pressed her lips together twice. The sound came out: "ma, ma."

Viki soon got the idea, and didn't speak until his fingers were in position against her lips; then looking up at him, she strained her lips forwards against his hand and said "mama" fiercely and with a thrust of the head when she was hungry. After a few days Dr. Hayes noticed her lips moving under his fingers, forming the word by themselves. He stopped helping her, though he kept his fingers in place. Gradually then he moved them around the side of the head until only the tip of his index finger remained, touching her upper lip. Finally he let go altogether, and Viki said "mama" without help. The lesson had been learned in two weeks.

Thereafter the chimpanzee said, or whispered, "mama" for anyone willing to reward her for doing so. It was noted that she preferred to touch her upper lip as she spoke, though she could utter the word perfectly well without this magic gesture, as she did when her hands were held.

Taking stock of their foster ape at the age of eighteen months, the Hayeses found it more illuminating to list her differences from, rather than her similarities to, Homo sapiens at the same stage of growth. For one thing she was hyperactive. All healthy children are active, mused Mrs. Hayes, but there was no comparison. Viki ran in four gaits on hands and feet, walked upright, jumped and leaped, and while she wasn't yet climbing trees, she had begun to show signs of soon doing so. "Her eyes seemed to be constantly glancing up, up, up, for a door jamb, a window casing, a human being, anything to climb."

The Hayeses told each other that exercise was all very well, but it left little time for intellectual pursuits, which were lagging anyway. At the stage where an average child uses about ten words, Viki was still stuck with "mama," and it was clear that she didn't always understand what was said to her. Her parents set out to teach her to imitate actions, with the intention of moving on from there to the imitation of sounds. Mrs.

Hayes would set her down to watch, then clap hands or blow a whistle, saying "Do this!" If Viki did, she was rewarded. At first, however, she did nothing but stare blankly. They put her hands through the motions, and then she seemed to understand, and imitated readily after that. She could copy six different actions—blow a whistle, pat the end of it as she blew, clap hands, pound on a can, draw a stick along a toy xylophone, and put a wooden bead on a string.

These stunts, explains Mrs. Hayes, were more than stunts: they came under the head of training. As Viki neared her second birthday the "imitation series" helped to add a second word to her vocabulary: "papa." It grew out of one of the noises she had learned by imitation to make, like a Bronx cheer. By manipulating her lips so that the series of *p*'s was slowed down, and calling at last for only two *p*'s in succession, the Hayeses elicited a perfect, if whispered, "papa." Viki used the two words indiscriminately and interchangeably as asking sounds.

She was just over two years old at the end of a busy summer, with much coming and going on the part of her parents. Mrs. Hayes took five days off from chimp teaching and went off for a vacation, and her reactions on her return are responsible for one of the most interesting sections of the book. Having come in late at night, she woke up in the morning with that sense of unreality that sometimes comes over those who have been travelling—Where was she? A small chimpanzee crept stealthily into the room and climbed to the foot of the bed to sit staring at her.

"She was a beautiful specimen with thick glossy hair and clear expressive eyes," wrote Mrs. Hayes. "I assumed that it was Viki, yet in a flash of panic I realized that I did not remember her at all!" The chimpanzee leaped to the dressing table and swung by a drapery cord—movements that seemed incongruous in a house—then swung back to the table and switched on the lamp. "This little act, performed by an ape's hand, also struck me as unreal and yet as strangely familiar."

The chimpanzee opened a perfume bottle, sniffed it, and screwed back the cap. She flicked her sideburns with a comb. She drew a hairbrush along her furry arm, and the room rapidly settled into a familiar pattern for her foster mother. Viki and Mrs. Hayes embraced, but there was still a strangeness: "Her hairy body, her big ears and eyebrow ridges seemed like some elaborate farce. I knew that this was my baby by her response to me, by the things she did with her hands, but someone had dressed my baby in a monkey costume." It took several days, she confessed, before there was no more embarrassment between Viki and herself.

Soon there was a setback in Viki's progress toward speech. She became estranged. She rejected Mrs. Hayes "with set lips and cold eyes." She ran away; she broke everything; she tore her clothing. "She stopped smiling completely," wrote Mrs. Hayes. "She refused to eat, refused to play, she forgot how to say her words." All day and all night she simply ran around with a wild look in her eyes, and Mrs. Hayes was frightened, wondering if this was to be the end of the experiment.

Thinking it over, she thought the trouble might have been the disturbed summer. Viki had been somewhat neglected, it was true, and her routine was often interrupted. Mrs. Hayes set out to make amends, slowly, carefully, and in the end successfully, wooing Viki back. "We had to retrain her to say 'mama' from the very beginning. We asked her to say 'ahhhh,' and then shaped the mouth for the m. But she took only one day to relearn it, and 'papa' came along with it. Not only that, but she was suddenly able to do all the various problems she had been able to do before the lapse."

Furthermore, she added another word to the list when she overheard Mrs. Hayes talking to Dr. Henry Nissen, the Yerkes Laboratories' assistant director. The ape already had a *k* sound and a *p* sound, said Mrs. Hayes, and they might be able to string them together to get a primitive version of "cup." Viki, overhearing the word, promptly said "k-p." Afterward she used the word "cup" more often than she used "mama" and "papa," and there was no doubt that she knew what it meant. Her thirst was always practically unquenchable: now she said "cup" perhaps a hundred times a day.

Six months after Viki's second birthday the Hayeses decided to take her on a motor tour to the North, stopping on the way at various speech clinics, psychology departments, and institutes for retarded children to seek advice. The speech specialists might have ideas as to how to deal with a child who had Viki's symptoms—only three words at the age of thirty months, words which she didn't always use appropriately even then. It had been hard to learn them, and she never used them in play as children do. "We had proved that her vocal apparatus was capable of speech, that she could imitate certain sounds, and that her general intelligence was quite adequate. What was she lacking, we could ask . . . and where do we go from here?" At the institutions for the mentally retarded they might learn of some type of child with Viki's shortcomings—lack of language and hyperactivity. In the matter of intelligence testing, there was the possibility that Viki wouldn't cooperate with a strange tester, but if she did and performed well on unfamiliar tests, her parents would have new confidence in the results. As Viki didn't always perform on request, Dr. Hayes proposed taking with them a sound film of her vocalizing, just to make sure. First they tried to record her ordinary chimpanzee reflex vocalizations. They tied up Viki and walked away, waiting for the usual piercing screams: instead she began crying "mama!" They gave up on that. Next they offered her a cup of milk, hoping to elicit the food bark, but Viki only said "cup!" She had become too civilized to make chimp noises.

The trip was a great success, except that the experts of whom the Hayeses had hoped so much turned out to be more puzzled than helpful. At Teachers College in New York they bombarded Mrs. Hayes with questions of their own, but their prognosis when at last she got it was encouraging. "They advised against pushing Viki. Let her tie down these three words before you confuse her with more, they said. They patted me on the back, said you're on the right track, and the meeting was over, leaving me happy, but not much wiser." At a training school for retarded children they were unsuccessful in their search for techniques for training and discipline. None of the children there behaved much like Viki. Those with her degree of language trouble were extremely retarded in other fields, including motor skills: whereas a child of six whose all-round intelligence was at the level of a normal three-year-old was still well ahead of Viki in speech and understanding. As to finding out how to cope with her exuberant personality, the children at that school were as gentle and well-mannered as most normal children. But they tested the ape on their Social Maturity Scale, covering self care, locomotion, play, and communication. Viki was judged to be eight months ahead of the normal human her age, though her total score was dragged down by that lagging language development, which was equal to a one-year-old's.

When all this was added up, the answer seemed to be that what the Hayeses had was a chimpanzee, not a child.

On the occasion of Viki's third birthday Mrs. Hayes made some observations, thus: Intelligence consists of three components—innate capacity, personal experience, and group intelligence. Group intelligence is preserved or collected through language—oral teaching, books, and daily communication with others. "Language conveys not only information, but ideals, traditions, and abstract philosophy. It fosters that cumulative thing—invention." No one man, isolated from birth, could cope very well with change in our environment: it is man's collective brain that has made us increasingly versatile, as it is man's unique ability to communicate knowledge that has led to civilization. Since apes do not acquire language to any significant extent—not even privileged apes like Viki—it seems unlikely that they ever have developed or ever will develop a civilization.

Not being a prophet, Mrs. Hayes was in no position to tell us when she wrote the book that Viki was to die at the age of six and a half. By that time she had learned, or was on the way to learning, the fourth word in her vocabulary, "up."

Later the Hayeses were divorced. A friend, commenting on this happening, said, "We used to think Viki was driving them apart—ruining the marriage, you know, but it looks now as though the chimpanzee was all that kept them together."

12

"You me go out please"

For some years nothing further in the field of chimpanzee communication was recorded. It was as if all the experts had grown disenchanted with the subject. Suddenly there was a small flurry: several new experiments were written up and reported at much the same time. The most recent news story of the sort, in the late summer of 1970, concerns a seven-year-old chimp named Sarah who since 1968 has been trained in Santa Barbara by David Premack, a psychologist. Sarah can express words, even simple ideas, with the aid of symbols—cut-out shapes of different colors, such as a blue triangle that stands for apple, and a red square that means banana. The cut-outs, mounted on steel, can be stuck to a magnetic board on the same principle that is used in certain children's toys. When Sarah showed that she understood which shape was which, Premack taught her to use another symbol for the preposition "on": thus she was able to say by this means, "Green goes on blue." She was then started on verbs. Her tutor is confident that she not only remembers the symbols properly but has grasped the extended meaning of them; she knows what redness means, recognizing that it is a color and also realizing—and stating—that a persimmon, like an apple, is red.

Those scientists—if there were any, which is doubtful—who still hoped to teach some super-Viki to talk, like a human being, with voice and tongue, were probably discouraged for good by the work of Dr. Philip Lieberman, professor of linguistics and electrical engineering at the University of Connecticut in Storrs. He is also on the research staff of the Haskins Laboratories of New Haven, a nonprofit organization that specializes in the fascinatingly varied studies of marine ecology, microbiology, and speech research. Dr. Lieberman's combination of subjects itself seems ill-matched until one remembers that there are many areas in which electronics and linguistics are closely allied. He is particularly interested in the difficulties encountered by retarded children in speaking. As he had told me in several conversations, he got on to the question of talking—or rather, nontalking—chimpanzees through observing Mongoloids, or as they are more technically known, victims of Down's syndrome.

"It was the shape of their lower jaws," he said. "Down's syndrome patients retain the lower-skull formation of infancy. The only way the skull can grow is up; that's why you get so many of them with big round heads."

But this is anticipating the complicated connections that brought him to consider such chimps as Viki. During 1968 and 1969, Dr. Lieberman with a number of colleagues published several papers: one was entitled *Vocal Tract Limitations on the Vowel Repertoire of Rhesus Monkey and Other Nonhuman Primates,* and another, *Newborn Infant Cry and Nonhuman-Primate Vocalizations*. Boiled down to words an untrained reader like myself can understand, the meaning of the work—or at least part of the meaning—was that apes and monkeys cannot talk because their vocal tracts, or equipment, are not made for speech. A chimpanzee's vocal apparatus (to select the subject that interests me most) is similar in important respects to that of a neonate, or newborn, human being. At birth a human's inner arrangements—at least, those back of the mouth—are arranged in a slightly different way than they will be later on, in that the larynx, which is the upper part of the windpipe (trachea) and is sometimes called the voice box because it contains the vocal cords, is higher under the head than it will be later when the baby grows older. At that stage it is only slightly below the pharynx, that part of the alimentary canal that connects the mouth with the gullet. The pharynx, to quote the dictionary, is the place into which nostrils, esophagus (gullet), and trachea all open. It is Dr. Lieberman's theory that this infantile closeness between larynx and pharynx makes it impossible for a newborn baby to form and utter words.

Very much the same arrangement of infantile vocal apparatus is found in newborn subhuman primates, with this difference: as the subhuman grows older, the relation between pharynx and larynx does not change as it does in the human.

Dr. Lieberman and his associates (in *Newborn Infant Cry,* etc.) recorded the cries of twenty newborn human babies from birth to the fourth day of their lives, setting up the necessary apparatus just outside a hospital delivery room and the nursery. Then they analyzed a sample of the cries, which included birth cries, fussing cries, angry cries, gurgles, hunger cries, shrieks, and inspiratory whistles. "Most of the cries were spontaneous," the report says, honestly but in somewhat sinister fashion, "some were elicited by pinches." Sound spectrograms showed that the cries were similar to the vocalizations of mature nonhuman primates. "Under certain conditions the laryngeal excitation was breathy. . . . The infants did not produce the range of sounds typical of adult human speech. This inability appears to reflect, in part, limitations imposed by the neonatal vocal apparatus, which resembles the non-human primate vocal tract insofar as it appears to be inherently incapable of producing the full range of human speech."

In *Vocal Tract Limitations,* etc., the Lieberman team discussed the vowel repertoire of a rhesus whose voice they analyzed in similar fashion to that of the newborns. They found that the monkey's vowels, compared with those of a human being mature enough to talk, indicated that the acoustic "vowel space" of a rhesus monkey is quite restricted, and that the limitation was due to the lack of a pharyngeal region that could change it. In conclusion, the report said, "These animals thus lack the output mechanism that is necessary for the production of human speech. Man's speech output mechanism is apparently species specific." This is a rather mystifying way of saying that man is the only animal that can talk, but scientists are like that.

Appealed to for elucidation of this passage, Dr. Lieberman showed me a medical drawing, one of those cross-section designs in which the head is cut straight through the middle, so that one looks at it from the side. It represented the vocal tracts of three primates: man, orang-utan, and capuchin monkey. In the two subhuman primates, he pointed out, the windpipe and gullet slanted in from the back on their upward trek to the mouth, whereas in man they stood up straight in a vertical position before turning a right angle into the mouth. "That position of the human windpipe and gullet is only found in a mature person," he said. "In the neonate the arrangement is like that of the monkey. You might find it easier to understand if you think of speech in terms of its two components—the source and the filter. The larynx is the source of the vowels, and the supralaryngeal vocal tract is the filter. Consonants are produced by constrictions in the vocal tract or by movements of the tongue and lips working on the vowels as they are produced by the source. But never mind all that; just keep in mind that a subhuman primate's vocal tract, like a newborn baby's, isn't fitted for speech, so the subhuman simply cannot talk."

Early in the academic term of Autumn 1970, Dr. Lieberman telephoned me from Storrs to say that his research into speech in the primate world had taken a giant step forward since he had made the acquaintance of Dr. Edmund S. Crelin, professor of anatomy in the Yale University School of Medicine, and that he would explain further, if I liked, if I met him there. I agreed, and went in search of Dr. Crelin's office in the Sterling Hall of Medicine on the Yale campus. Finding a door with the right number and a sign that told me to walk in, I obeyed and found myself in a large laboratory, of which the most noticeable furniture was a table on which were arranged in three rows a number of skulls or skull casts, in various shapes, sizes, and colors. No one else was there, but through an inner open door I saw an office with Dr. Lieberman and another man in it, deep in enthusiastic talk.

Dr. Lieberman hailed me and asked me in; he introduced me to Dr. Crelin. He said, "I spent the summer in Europe, looking at primitive primate skulls. It all fits in awfully well with what we were discussing the last time we met, and Dr. Crelin agrees. It was amazing that he and I should have met when we did, just when each of us knew something to help out the other. Things happen like that sometimes. To begin with, he's recently published this book." He handed me a large, handsome, blue volume with the title, *The Anatomy of the Newborn.* "It's exactly what I needed," he said happily.

Dr. Crelin, discounting credit, said, "I happened to be the first to think of doing a book like it—that's all. Anatomists needed something like this, but they didn't realize it: all these years nobody's thought of making a special study of the newborn, and I don't know why. Obviously it was wanted. Everybody's asking for it now. I give a course here in developmental physiology, and I know. You see, all these years people have been treating the newborn like a miniature adult, which it isn't by any means. All you have to do is look. See here." He indicated a drawing of a baby's skeleton. "Look at it," he urged me. "See that head? See the proportions of the limbs? Nothing like an adult, is it?"

I admitted that the skeleton had very strange features—long arms, short legs, and a very big head, though the head had scarcely any lower jaw. It looked distorted.

"The skull is what interests us," put in Dr. Lieberman. "It ought to remind you of something else. Are you familiar with Neanderthal man?"

"Well, not really," I said apologetically.

"Never mind," said Dr. Crelin. "We'll introduce you to him in a minute." He took up a lecturing stance. "As you probably know, the human fetus goes through evolutionary stages. For a while we have gills, and flippers, and a tail and so on. It's the old saying, you know—ontogeny recapitulates phylogeny. It's never been said better." Perhaps not, but later I found a translation that I think is easier: The development of the individual reflects that of the species.

"Haeckel's not so bad after all," continued Dr. Crelin. "I know it's the fashion nowadays to neglect him, but he was right. Anyway, we go through all these evolutionary stages, and some of them aren't complete by the time we're born. That's why the neonate skull looks distorted: it hasn't finished evolving. And that's why a newborn human represents a sort of subhuman primate."

"In a newborn, the larynx and the pharynx—that is, the windpipe and the top of the alimentary canal—run like *this,"* said Dr. Lieberman, making a sketch to show me. "They are independent of each other. That's why a baby can suck and breathe at the same time. You or I couldn't do that: if by mistake we got a bit of food lodged in the trachea—the windpipe—we would probably choke to death. It happens to people sometimes. A newborn doesn't run that risk: the functions of each channel don't interfere with the other. So you see we pay a price for the ability to speak. Now let's look at Dr. Crelin's skulls."

We all moved into the laboratory, to the table where the casts were arranged. "I'll tell you in advance that Neanderthal man couldn't talk," he continued, "and the reason we know it is that he was like the newborn today: he didn't have the right equipment."

Dr. Crelin said, "We use the term Neanderthal for all types in that period—the classic type, the man of the last glacial period. As you can see here, there's a wide selection, but this fellow here, La Chapelle man, is the one anthropologists look on as one of our direct ancestors. I'm not so sure. You've probably heard it said that Neanderthal man is so much like us, if he came into a subway train nobody would take a second look at him. Well, I can tell you, if he got into *my* train, *I'd* notice him all right."

We stood looking at the array of skulls. Dr. Crelin picked up a small one at the far end. "Here's our newborn," he said. "Here's a young chimp. Here's the Taung baby. Look at all the differences. Anthropologists have always classified these things by the frontal skull—you know, beetling brows and all that—but I think it's not as significant as what we're on to, the naso-pharynx." He turned the baby's skull upside down to illustrate. "Here. See that hollow? It's not as deep as it will be in an older human, but it's as deep as a chimpanzee's head will ever show it.

That's why I have my doubts about La Chapelle man. Here we have a fully grown adult, but look at what it's like." He turned the skull over. "Just like Newborn, do you see?"

"In fact, La Chapelle man is just a big baby," said Dr. Lieberman.

Dr. Crelin said, "That's why I have my doubts about this fellow." He replaced the Neanderthal skull and patted the top of its head. "I don't think he's one of our ancestors: I think he went off on a tangent, and evolution went round somewhere else—like this, perhaps." His gesture took in another lot of skulls. "I think the chimpanzee is just as much on the right track as Neanderthal," he added. "Just as closely related."

"Neither Newborn nor Neanderthal have a chin," said Dr. Lieberman. "And just look at this mandible. . . ."

The two doctors entered into a fascinating dialogue about the possibilities of other prehistoric ancestors to man. Dr. Lieberman thought that the transitional period might be represented by Piedmont man—something between Neanderthal and modern. Dr. Crelin wondered if there might not have been some amount of crossbreeding between these fellows, as he called them. "There's no doubt," he said, "that Cro-Magnon *could* speak."

"Crossbreeding is certainly possible," Dr. Lieberman said thoughtfully, "but you've got to remember the linguistic boundaries. There are birds of the same species who have different calls, and they don't usually crossbreed. The same thing might be true of these people. Those who could speak might not have anything to do with those who couldn't."

"Couldn't Neanderthal make any noise at all?" I asked. It was the right question, evidently. Dr. Lieberman handed me a paper he had recently written with Dr. Crelin on just that subject: *On the Speech of Neanderthal Man*. I began to look through the pages until I came to a passage that seemed pertinent:

> In order to reconstruct the supralaryngeal vocal tact of Neanderthal, it was essential to locate the larynx properly. Because of the many similarities of the base of the skull and the mandible between Newborn and Neanderthal, coupled with the known detailed anatomy of Newborn, of adult Man and of apes, it was possible to do this with a high degree of confidence. Although the larynx was judged to be as high in position as that in Newborn and apes, it was purposely dropped to a slightly lower level to give Neanderthal every possible advantage in his ability to speak.
>
> Once the position of the larynx in Neanderthal was determined, it was a rather straightforward process to reconstruct his tongue and pharyngeal musculature. . . .

I broke off and looked at Dr. Lieberman with awe. "Do you mean you've *constructed* a talking Neanderthal man?" I demanded.

"Why, of course. We had enough data," he said. "With computers and all, it was simple. He made noises all right, but they're unformulated—like baby's cries. Or rather, like this." He suddenly made a noise hard to describe—rather like a bark but not quite. It was quite a loud noise, "Ugg! Ugh! Ough!" Just then the outside door of the laboratory opened, and a female student peered in, looking frightened.

"All right," called Dr. Crelin to the puzzled child. "I'll be free in about a quarter of an hour. Come back then."

The door swung shut. "Of *course,* we know how Neanderthal sounded," said Dr. Lieberman. "If you like, I'll send you a tape of it."

At the end of a chapter on "Anthropoid Speech and Its Significance" in *Almost Human,* Yerkes says, "I am inclined to conclude from the various evidences that the great apes have plenty to talk about, but no gift for the use of sounds to represent individual, as contrasted with racial, feelings or ideas. Perhaps they can be taught to use their fingers, somewhat as does the deaf and dumb person, and thus helped to acquire a simple, nonvocal, 'sign language.' "

I was reading the October 25, 1968, issue of *Science* when memory of that paragraph stirred in my mind, and at the same time I felt a pricking of the thumbs. The article that brought on these sensations was written by our old friend Winthrop N. Kellogg, now professor emeritus of experimental psychology at Florida State. Under the title "Communication and Language in the Home-Raised Chimpanzee," Professor Kellogg summarized the literature accumulated to date on the speech, or nonspeech, of apes in the home, with various comments made by observers on gesturing rather than speech as a means of communication. He repeated the story of Gua and the bottle, and mentioned Viki's way of taking Mrs. Hayes's hand and leading her wherever Viki wanted to go. He recalled that the orang Dr. Yerkes once worked with in California was described as doing the same thing, taking Yerkes' hand and pulling. Mrs. Kots of Moscow was quoted as saying that her chimpanzee Joni made gestures rather like those made by her infant son, requesting with outstretched hand, rejecting food by turning his face and head away, indicating thirst by putting his hand to his mouth, drawing attention to himself by tugging at her dress. The point of all this is that two people in Reno, Professor R. Allen Gardner and his wife Dr. Beatrice T. Gardner, both of the University of Nevada, have been trying out the idea expressed in that paragraph of Yerkes'. For various reasons the experiment has been kept fairly private, but the Gardners have confided in brother psychologists, including Professor Kellogg.

In June 1966 they acquired a wild-born female chimpanzee about twelve or eighteen months old—they could not be absolutely sure of her age—and set to work. Washoe, the chimpanzee, lives in a trailer in their yard, and her existence is governed—theoretically—by hard-and-fast rules. She spends all her waking hours in the company of one or more humans who address her, and each other in her presence, not vocally but by means of the American Sign Language for the deaf and dumb, widely used in North America. Only a genuine sign language like ASL would have answered the purpose: those others that depend on the spelling-out of words are no use, since Washoe is illiterate. For the experiment the Gardner team all had to learn ASL. Ordinary speech in her presence is taboo, for fear it might confuse her.

The experiment had begun to pay off when Professor Kellogg wrote his article, in the late summer of 1968: by that time he was able to record a number of the signs she made. She beckoned, using either wrist or knuckles as a pivot, to say either "Give me" or "Come." When she wanted to be lifted, perhaps placed on someone's shoulders to reach an object, she said, without words, "Up," by pointing skyward with her index finger. A loud or strange sound caused her to touch her ear with the index finger, for "Listen!" To plead, as if saying, "Please," she rubbed her chest with open hand, then stretched out the hand in the "Give me" gesture. For "Hurry up!" she shook her open hand at the wrist—a fair approximation of the ASL gesture, though a human would keep his index and second fingers together. (Washoe's hurry-up gesture is not unlike one made by chimps in the wild for the same purpose.) In sum, not bad for a three- or three-and-a-half-year-old chimpanzee.

The *Science* article excited me, and started me looking for people who knew something more about Washoe. Dr. Philip Ogilvie, then director of the Oklahoma City Zoo and professor of biology, was able to help: he wrote to me that his friend Professor William B. Lemmon of Norman was also a friend of the Gardners and had himself made the acquaintance of Washoe. Professor Lemmon had returned from a visit to Washoe and reported to Dr. Ogilvie that what the latter called "this sign-language thing" really works. Very kindly and at long distance he tried to fix up a meeting with the Gardners for me, but they shied off. It was just the wrong time for publicity, they explained—they were coming to the end of the money they had received as a grant for their research, and now it looked as if Washington would turn down their application for a new one. The year 1969 bade fair to be a gloomy period altogether for scientists: Washington was turning down requests for money with monotonous regularity. The Gardners, in the middle of a fascinating and costly experiment, were full of apprehension: everything might have to come to an end before the middle of the year. And then, too, Professor Gardner explained, Washoe needed isolation. Too much company, too much diversion, spoiled her powers of concentration. I gathered from Philip Ogilvie on the telephone that she was beginning to behave intractably most of the time. "Bill says she's spoiled," he said.

Spoiled or not, I wanted to know more about her, so I was very pleased to hear that the Gardners were coming to Washington to give a lecture to some of their peers on the subject of Washoe and the sign language. Because they felt that the chimpanzee herself must not be subjected to the stresses of travel and public appearances, she would remain behind in Reno, but they were bringing films to illustrate their lecture. I, too, promptly went to Washington.

It was early afternoon, and the hall was very full of people one might expect to show interest in a sign-language ape—psychologists and linguists mostly. Beard incidence was abnormally high. Many people had brought their children. I could imagine a typical conversation as heard that morning at local breakfast tables: "Billy, how would you like to see a chimp show? Okay, pick me up at the office." In the crowd I couldn't see anyone who might have been the Gardners until they got up on the platform.

They were a slender couple. Professor Gardner looks like a youngish professor, which is what he is; and Dr. Gardner is a good-looking woman with dark hair. She gave what I might call the prologue—some introductory remarks on what the American Sign Language is and why it is being used. Washoe, she said, was three and a half or four now, which is still very young for a chimpanzee: some of these apes have been known to achieve the age of forty and even more. She would become a full-fledged adult at fourteen, but would be sexually mature at seven or eight. For the past thirty-two months she had been exposed only to the American Sign Language as a means of communication. By now, said Dr. Gardner, she was using over seventy gestures, though the films we were waiting to see, having been made earlier, exhibited the use of only thirty. Dr. Gardner warned us to watch the film closely and be alert, as Washoe sometimes talked fast. The effect, she said, is something like a foreign language spoken rather quickly by a child. I took note of a few random remarks she made as she wound up:

"In playing games, Washoe makes much use of pronouns, thus—I play, you play.

"She speaks ASL in her own way, because it is not always easy for her to make the absolutely correct gesture. However, one learns to understand what she is saying—" as if, I reflected, she was talking with an accent that one must automatically discount.

Dr. Gardner then signalled; the lights went out; and the film went on. In the dark her voice continued in commentary. We saw a small chimpanzee in singlet and diaper standing outside a back door, looking into the camera. One of her hands was held out pleadingly on an almost straight arm. "Washoe saying 'Come' to a friend," said Dr. Gardner. A tall youth came into the picture and took the ape's hand, after which she turned and held out her other arm in the same gesture. " 'You come, too,' to another friend," interpreted Dr. Gardner. In the picture, she too walked onstage to take Washoe's other hand, and the trio marched off scene.

Then we saw the chimpanzee sitting on a couch, being playfully hit by some invisible actor with a pillow. Washoe laughed and rolled with the punch. Another pillow softly hit her, and then nothing happened until she pushed her doubled-up hands, or fists, together over her head, two or three times butting them against each other. "That's baby talk for 'More, more,' " said the interpreter. "The proper gesture is in front of the chest, but Washoe always does it over her head."

In the next scene Washoe was riding piggyback on Dr. Gardner, making big swoops down through the air with her right hand. "Washoese for 'Go,' " explained the real Dr. Gardner. "Now I ask 'Go where?' " In the picture her hands, as if joined at the wrist, spread out fanwise in a horizontal plane. The fingertips of Washoe's two hands came together and separated. "She is saying 'Out there,' " explained Dr. Gardner. "Now you see her running to a closed door. See, she's shaking her hand at the wrist and then making the opening sign. In other words, 'Quick, open.' . . . As you see, she's quite legible, and we don't press for finer diction for fear of temper tantrums. They would interrupt the work pretty drastically."

Washoe on the screen hugged herself, demonstrating love for a human companion. She begged—"Please. Please." Sitting by a large carpetbag from which an attendant brought out one object after another, she showed what each one was: hat, dog, pants, baby (doll), shirt, shoes, brush, comb. It was true; she did talk fast. For "oil" she rubbed herself. Brushing her teeth, lips drawn back, was "toothbrush." For that all-important word "banana" she peeled an imaginary one. (I felt a sudden wave of regret that Dr. Yerkes would never see Washoe say "banana.") She even said sentences, like "The flower is sweet." She indicated herself by touching both ears with her index fingers.

"That's her name," interpolated Dr. Gardner: "Washoe Big-ears. Now you see her climbing a tree. Look carefully—she's just seen a bird, and is telling us about it." Washoe, hanging by one hand from high up in the tree, was signalling "bird" to the watchers beneath, with fingers rapidly closed and opened.

Again on the ground, she looked at pictures of objects in a book and translated them into her form of words. "She used to try to pick the things up, especially the fruit," said Dr. Gardner, "but she knows better now. There, if you watch, you can see what she's saying: 'bib,' 'hat,' 'pants,' 'clothes,' 'cat.' " For "cat" Washoe inscribed whiskers on her own face, and for "dog" patted her skinny little thighs.

At this point something happened to the projector and the films had to cease: thus we missed the latest pictures of Washoe. The lights went on, and Professor Gardner replaced his wife on the dais. There wasn't much to add, he said—the pictures told the story.

He did, however, add a good deal. He said that we should not assume from the appearance of the film that their subject always went through her paces so promptly, one after another. Interruption was the constant pattern. "It was hard enough getting her to sit in one place for long enough for one or two words," he said earnestly. "As a result we often didn't catch something especially interesting, as when she would say 'A dog is barking' suddenly, on her own."

Soon enough, he said, one lost that awed feeling about the business, working with Washoe. At first they had worried about how they could induce her to transfer an idea from one situation to another. "But it didn't matter," continued Professor Gardner. "Washoe always transferred on her own. For instance, she is seldom near real dogs or cats, yet it's no problem for her to recognize toys or pictures of dogs and cats. Just as she's hardly ever been close to a real baby, but she knows about babies: in her vocabulary 'doll' and 'baby' are the same thing."

He told us that they had once taken Washoe to a school for deaf-and-dumb children, where "the children read Washoe perfectly." He was carrying the chimpanzee piggyback when they met another man carrying a small deaf-and-dumb child. Washoe saw the child and said, or gestured, "Baby!" and at the same moment the child said, "Monkey!"

Idea transference sometimes made Washoe hard to handle, he said. Take keys, for instance. (At this point, as may be imagined, I sat up and listened harder than ever.) Washoe learned to use keys in padlocks and seems to have mastered this technique for all time. For two months after she demonstrated that she understood it, all padlocks and their corresponding keys were kept away from her, but when she did get her hands on a key again, she knew exactly what to do with it. She would try all the keys she could get hold of on all the locks she could reach. Now, asked Professor Gardner, why should a gesture transfer with Washoe to all classes? Simply because it does, he replied to himself. Most animals do transfer. Why, then, do *we* think it difficult, when it is so easy for them?

"Psychology," he said, "is the relation of behavior to experience. You can't be a psychologist unless you're a behaviorist of some kind." The audience looked thoughtful, and some, I thought, didn't agree, but he went on: There were many statements or sentiments expressed by Washoe in sign language in which she seemed to find no difficulty in putting thoughts together. She said of the refrigerator, "Open food drink"—not, as one might have expected, "Open cold box," but her version was just as much of a combination. One of her longer sentences was, "Please come give me hurry more." She often said, "Please open hurry" and "Give drink please." She said, "Go in" and "Go out," and once, when she wanted to go over to a raspberry bush, "Go sweet." She said, "Roger come tickle" and "You me go out please."

"At first, all this is very exciting," Professor Gardner admitted, but he said that later one began thinking: Is it really so staggering? Obviously it opens new horizons in more than one way, but when you see it happen, it begins to lose its mystery. You think, Why not?

I sat trying not to find it exciting. I failed.

Professor Gardner continued, "A sharp distinction between semantics and syntax isn't easy for the psychologist to make, probably because the distinction is *not* sharp. Perhaps there actually is no line between them; there may be a kind of merging instead." Once more he had lost me. Semantics means meaning in language, and syntax is orderly arrangement of words, or sentence structure. He was saying—well, the sentence structure would of course affect the meaning, and the meaning would direct the structure. Yes, I saw what he meant, or thought I did. Logic and language are interdependent. . . .

But the professor had gone on. It seemed clear, he said, that gestures are the best way we have to communicate with chimpanzees, and he would be surprised if chimpanzees in the wild don't use gestures more than we have thought in the past. Some observers bear this out. Certainly Washoe is always looking for new gestures or words, and sometimes she even invents one, as she did in the case of "bib," about which he told this story: The Gardners don't claim to be absolutely proficient in the American Sign Language, and sometimes they made up a gesture for Washoe instead of tracking it down in the book. In this way they invented for "bib" the rubbing or wiping motion we had seen—hand over nose and mouth. But one day Washoe asked for her bib with a new gesture, using her index fingers to trace lines from her shoulders down to meet on her chest, in a kind of outline drawing of a bib. This was interesting, but after talking it over, the Gardners decided not to accept Washoe's ver-

sion. After all, explained the professor, the object of the exp
to teach her a language, not to learn one from her, and t
using the rubbing gesture. Washoe sometimes used theirs, so
own, until at a meal at the school for deaf-and-dumb childre
for her bib with the rubbing sign. A teacher said, "What's
Gardners told him: "bib." He shook his head. "No," he said. "The sign for bib is this"—he drew the outline of a bib with his index fingers, from shoulders to chest. "Now we're not sure what we're teaching her," confessed Professor Gardner. "Sometimes we use one, sometimes the other."

He talked about hieroglyphics, which are, he reminded us, pictures and nothing else, though it was necessary to find the key before we could read them. Communication by gestures between human and chimpanzee is much the same kind of thing. It might be asked, he said, why nobody has ever before tried to talk with chimpanzees by signs, but Professor Gardner thought it might be because so many people identify speech with language. Also, people have a low opinion of chimpanzees, having seen them in laboratories: they look on the apes as a sort of large, expensive, nasty rat. In fact, chimpanzees are far too intelligent to be easy or comfortable subjects: the problem is to give them enough interesting problems to keep them occupied and developing. There are very interesting things to do with chimpanzees, he said, if you let them decide for themselves what to do.

As to Washoe's future, Professor Gardner understandably had only a few generalized things to say. She will be lonely unless two or more teams apart from his can train other chimpanzees in the same way. Suppose some day in the future she has babies. Will she be able to teach them her new language? Possibly, even probably. She does push, even now, when her human friends don't seem to understand what she is saying. Deaf children who have met her say that she is better at talking than the Gardners give her credit for. And, of course, she is imitative, like chimpanzees generally, taking her imitations very seriously. "She even sews. It's very impressive," Professor Gardner assured us.

The combined, illustrated lecture was at an end, and the meeting was declared open for questions. These came thick and fast. Some were too technical for me to follow; others weren't. One man asked if Washoe ever talked in her sleep, and Professor Gardner said, "We don't know: we don't watch her then. We're so thankful when she goes to sleep that we feel we should let sleeping chimps lie."

Did she signal approval and negation? Yes, he said—in fact, she shook her head for "No." Another questioner wanted to know if she actually asked for signs, and he said, "We wish she would. She doesn't yet, but she is accelerating and we have hopes. One of the disappointments we've had so far is that *she* has never asked *us* a question. One of our workers says she doesn't have to, she questions with her eyes. But then, she's still young."

When I left the hall it was still thronged with psychologists in groups, vocalizing and vocalizing.

Afterward

Before I saw, heard, and read so much about primates in colonies and the laboratory, I took it for granted that I would be able to tidy up and put away the subject when I had written about it, as one does after reviewing a book or passing an examination. One puts the review or examination paper into an envelope, sends it off to the proper recipient, and turns to some new study. Well, I was wrong. I have not finished with monkeys and apes: I will never be finished. An awareness of primates is going to accompany me for the rest of my life. During the time I spent on the subject, I have acquired more understanding of my own species.

Not that we are so much alike, except here and there in details such as fleshly tissue, teeth, and resistance to this virus or susceptibility to that infection. I must keep in mind that we *aren't* all alike. To people with a talent for generalization, who believe, for example, that war can be thoroughly understood through studying the behavior of animals or fish defending their own territory, it is tempting to put faith in subhuman primates as models for human institutions—marriage, family, and government. The trouble with this concept is that among primates there are as many systems of family, mating, and dominance as there are primate species.

Which must we accept as the shining example of domestic peace and happiness—the promiscuous chimpanzee, the easygoing gorilla, or the monogamous gibbon? Which is the better mother, the clutchy pig-tailed macaque or the permissive stumptail? Japanese macaque fathers help to look after their babies: marmoset parents take it turn and turn about to carry their offspring, but chimp fathers seem to feel no responsibility whatever for theirs. In selecting a leader, would you prefer a bossy, overbearing type like the baboon—a totalitarian leader if ever there was one —or a gentle father figure like the gorilla? Questions like these are not easy for human primates to answer. Indeed, they are impossible to answer as far as I can see—and I can see a lot farther now than I could before I began observing primates. In all directions, horizons have receded. The glib summary no longer comes readily to my lips. I am sure of only one thing—that the proper study of mankind is more than man.

Selected Bibliography

The following books are suggested for further reading, with the reminder that things move fast in the world of primatology and no book on the subject, if published more than a few months ago, is fully up to date. Nevertheless, much that was written in the earlier days of the science should not be missed.

BROADHURST, P. L., *The Science of Animal Behavior*. London: Penguin Books Ltd., 1963.

DE VORE, IRVEN (editor), *Primate Behavior: Field Studies of Monkeys and Apes*. New York, Chicago, San Francisco, Toronto, London: Holt, Rinehart and Winston, 1965.

KÖHLER, WOLFGANG, *The Mentality of Apes*. New York: Vintage (Random House, Inc.), 1956; first published in English by Routledge and Kegan Paul Ltd., London, 1925.

LENNEBERG, ERIC H., *Biological Foundations of Language*. New York, London, Sydney: John Wiley and Sons, Inc., 1967.

MARAIS, EUGENE, *The Soul of the Ape*. London: Anthony Blond, 1969.

YERKES, ROBERT MEARNS and ADA WATTERSON, *The Great Apes*. New Haven: Yale University Press, 1929; London: Oxford University Press, 1929.

ZUCKERMAN, SOLLY, *The Social Life of Monkeys and Apes*. London: Paul, Trench, Trubman and Company Ltd., 1932.

Index

[Page numbers in italics refer to illustrations.]